STUDIES IN CONVECTION
Theory, Measurement and Applications

Volume 1

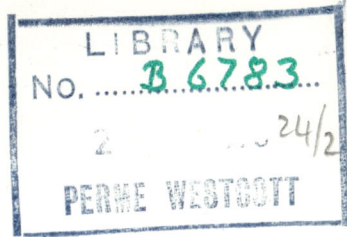

ROCKET PROPULSION ESTABLISHMENT LIBRARY

Please return this publication, or request a renewal, by the date stamped below.

Name	Date
R.St. PARKINSON	4.5.78

(4/64) L23964 442077 Wt29280 D7061 10/64 10M T&Co G871.   R.P.E. Form 243

STUDIES IN CONVECTION
Theory, Measurement and Applications

Volume 1

Edited by

B. E. LAUNDER

Imperial College of Science and Technology, London

1975

ACADEMIC PRESS
London   New York   San Francisco
A Subsidiary of Harcourt Brace Jovanovich, Publishers

ACADEMIC PRESS INC. (LONDON) LTD.
24/28 Oval Road,
London NW1 7DX

United States Edition published by
ACADEMIC PRESS INC.
111 Fifth Avenue
New York, New York 10003

Library of Congress Catalog Card Number: 75-13621
ISBN: 0 12 438001 8

PRINTED IN GREAT BRITAIN BY
J. W. ARROWSMITH LTD., BRISTOL

Preface

The present volume is the first of a series of
books that is planned to document new developments in
the field of convection processes. The aim of 'Studies
in Convection' is to provide an up-to-date presentation
of major contributions to the subject, with particular
reference to the development of methods of calculating
turbulent convection processes. There are a number of
monograph series already providing articles within this
field. These tend, however, to take as their remit a
breadth of subject matter that is much wider than simply
"convection" yet which commonly fails to encompass *all*
convection processes: a series in "Fluid Mechanics"
may not contain articles on species transport while one
in "Heat and Mass Transfer" may limit attention to non-
reacting flows. Perhaps, largely for this reason, while
no self-respecting institutional library would be without a set, relatively few volumes of this kind find
their way into personal collections. By limiting attention squarely to convection processes it is hoped to
provide volumes that will prove of value equally to the
active research worker in convection and to the new
graduate student in the field.

Each volume will contain four or five extensive
articles on a single aspect of convection. All the contributions will be "survey" articles but under this um-
brella several variants may be found. These variations
can readily be discerned in the contributions to the
present volume. Dr. Rodi's article provides, at least
in form, an orthodox review of isothermal, free turbulent shear flows. It is comprehensive and properly
critical, its author's conclusions being arrived at from

extensive checking of the internal consistency of the
published data. The articles by Professors Whitelaw
and Patankar are concerned with topics which, though
not strictly new, are none the less now in a rapid
state of development; each lead teams that have contributed to the advancement of their subject. Here,
therefore, I have asked them to provide a personal
view. Professor Whitelaw's article traces the contributions of his group to the various aspects of laser-
Doppler anemometry over the past five years. Information is provided on such topics as signal processing,
optical units, particles and laser power; more than
60 references are cited providing further details of
these subjects. Professor Patankar provides a development of his three-dimensional finite-difference procedures for convective flows; he manages to provide a
comprehensive account without recourse to the intermi-
nable sequences of equations that usually characterize
papers on such topics. A number of computational results are presented which serve admirably to underline
the exciting prospects for work in three-dimensional
flow over the next few years.

        The final article in the volume is again a personal
view, but of a rather different kind. Professors Kays
and Moffat, in looking back over their ten years' work
on transpired boundary layers, have taken the oppor-
tunity to make a thorough re-evaluation of their group's
work. With the benefit of their accumulated experience,
the experimental results have been arranged to bring
out more decisively than in the original works the struc
tural effects of pressure gradient and mass transfer on
the flow development. Here, as in the other articles
in the volume, the authors' aim has been to provide a

survey of the state of knowledge in the field that
will be an appreciated companion of newcomer and ex-
pert alike.

In producing the volume my editorial task has
been much lightened by constructive comments on individual articles by experts from around the world. I
would record also my thanks to the authors for their
helpfulness and patience in providing articles in har-
mony with the overall aims of the series. Finally my
thanks are due to Margaret Hudgell who, despite
scarcely legible and sometimes inconsistent editorial
scribbles, typed the volume with her customary precision.

Kensington   BEL
February 1975

Contents

NUMERICAL PREDICTION OF THREE-DIMENSIONAL FLOWS

by

S.V. Patankar

Department of Mechanical Engineering
University of Minnesota, Minneapolis, Minnesota, USA

ABSTRACT

This article is a compilation of some recent work towards the development of a general prediction procedure for three-dimensional flows. A general calculation scheme is first described for elliptic flows; and then its simplified version for parabolic flows is presented. The scheme is an implicit finite-difference method, in which the difference equations are formed by integration over a small control volume surrounding a grid point. A hybrid formula, which is a combination of the central- and upwind-difference schemes, is employed to represent the convection and diffusion terms. The flow field is characterised by the three velocity components and the pressure. Their calculation proceeds in two steps: first, an estimated pressure field is inserted in the momentum equations to obtain a preliminary velocity field; then appropriate corrections to the pressure field are calculated such that the resulting velocity field will satisfy the continuity equation. Mathematical models for the physical processes of turbulence, combustion and radiation are outlined, and their use is illustrated by some computations of three-dimensional flows. For the results of three-dimensional parabolic flows, comparisons of the predictions are made with experimental data; the agreement is satisfactory.

1. INTRODUCTION

1.1 *Purpose of the Present Article*

Fluid flows occur in the human body, in and around engineering equipment, and in the environment. In each of these fields, it is of practical importance to know quantitatively how the flow will behave under a given set of conditions. A new design of engineering equipment should ideally be based on an accurate prediction of how the equipment will perform. Operation of existing equipment could also be optimized if the quantitative influence of various operating conditions could be determined. There is much economic advantage in being able to predict and perhaps to control the effects of floods, winds and fires.

There are two main ways in which the required quantitative information is obtained. The first is the experimental approach, while the second involves a theoretical calculation. The experimental path is more popular, especially in the design of engineering equipment. But this path is usually very expensive; and, when the size of the actual equipment is large, the designer is forced to draw his conclusions from the tests on scaled-down models of the equipment which do not duplicate all the important factors.

The path of obtaining quantitative information through a theoretical calculation is, in principle, more satisfactory. When fully developed, it will give predictions which are accurate and inexpensive to obtain. Research in recent years, coupled with the availability of large and fast computers, indicates that reliable and flexible mathematical models for most fluid-flow phenomena of practical interest should indeed be available in a few years' time. Already it is possible to replace much of the model testing in the laboratory by a theoretical calculation.

Theoretical methods are often classified into two groups: analytical and numerical. The former class is so severely limited in its applicability that it immediately follows that a general calculation method has to be of the numerical variety. Further, since it is the solution of partial differential equations that largely concerns us, the requirements of generality narrow our choice to finite-difference methods. (The term "finite-difference method" is used here to indicate a method in which attention is focused on the values of variables at a finite number of points; the so-called finite-element method also can be considered to belong to this type.)

A theoretical prediction procedure consists of two components: the finite-difference solution procedure, and the mathematical description of various physical processes such as turbulence, chemical reaction, etc.. The recent advances in our prediction capability have been the result of the progress on both fronts: some good numerical algorithms have become available, and sufficiently general mathematical models of turbulence and of other physical processes have been devised.

In recent years, the author has been associated with the work led by Professor D.B. Spalding at Imperial College, London on the calculation of three-dimensional flows. The purpose of the present article is to describe the basic numerical method

used in this work and to show some examples of its
use.   Although three-dimensional flows are given
almost exclusive attention, a calculation procedure
for two-dimensional flows can be easily derived as a
special case.   Models for processes like turbulence
are not given full attention in this article, but
illustrative material is included to indicate the
present practices.   In general, the article is not
designed to enable the reader to start writing a computer program;   the aim is rather to give him an over-
view of the present state of the art.

1.2   *Outline of the Article*

Historically, the method described in this article
grew as an extension of some work on three-dimensional
duct flows.   These flows, although three-dimensional,
can be predicted by a marching procedure which essentially solves a two-dimensional problem at each step
in the marching process.   These flows, named as parabolic flows, can be considered as a special case of
general three-dimensional flows to which the marching
technique may not in general be applicable.

The present article is therefore divided into two
parts: Part I presents the general case and Part II
deals with the special class of flows which can be
solved by marching in one space co-ordinate.

Part I covers Sections 2-5.   Section 2 describes
the type of flows under consideration and comments on
the available methods of their prediction.   Section
3 is devoted to the description of the calculation
procedure with which the author is associated.   Here
an attempt is made to outline the basic ideas of for-
mulating and solving the finite-difference equations.

The method uses the velocity components and the pressure as the hydrodynamic variables; their linkage is handled in a special way which is described in Sec. 3.5.

Some of the algebraic details of the mathematical models for turbulence, combustion and radiation are given in Sec. 4. These details serve to show the reader the kind of information a general calculation procedure needs; and some of these details form the basis of the computations to be presented in the remainder of the article.

Section 5 contains five examples of three-dimensional-flow calculations of varying complexity. Of special interest are: (a) the flows with distributed resistances such as the flow in the shell-and-tube heat exchanger, and (b) the flow in a gas-turbine combustor in which simultaneous effects of turbulence, chemical reaction and radiation are taken into account.

Sections 6, 7 and 8 form Part II of the article. Section 6 defines and illustrates the term "parabolic flow". A review of existing calculation procedures for this class of flows is contained in Sec. 6.2. The adaptation of our general numerical procedure to steady parabolic flows is described in Sec. 7. The essential trick there is the decoupling of pressures that are used in the longitudinal and cross-stream momentum equations. Examples of computations of turbulent parabolic flows are given in Sec. 8; the results are compared with experimental data wherever possible.

Concluding remarks are made in Sec. 9.

PART I.   THREE-DIMENSIONAL ELLIPTIC FLOWS

2.   ELLIPTIC FLOWS AND THEIR CALCULATION

2.1   *The Problems Considered*

It is customary to classify fluid-flow problems as: (a) one-, two- or three-dimensional, (b) steady or unsteady, and (c) governed by parabolic, elliptic or hyperbolic differential equations.   Until now most theoretical work has been confined to two-dimensional flows.   Yet, practically-important flows are often three-dimensional.   Plane two-dimensional flows are contrived only in laboratories and there, too, three-dimensional effects are usually detectable. Cylindrical geometries have greater promise of creating a two-dimensional (i.e. axi-symmetrical) flow, and some practical situations can indeed be reasonably thought of as axially symmetric.   But often certain conditions at the boundaries destroy the axial symmetry; for example, the flow in an annular gas-turbine combustor does not remain axi-symmetric when the fuel is injected through a number of discrete holes rather than through an annular slot. Rectangular and other non-circular geometries almost always give rise to three-dimensional flows.

Thus it can be concluded that the aim of being able to predict all practically-important flows can be attained only through a three-dimensional-flow calculation procedure.

Three-dimensional flows can be classified as parabolic or elliptic depending on the types of the differential equations that govern them.   (For compressible flows, the equations become hyperbolic under

certain conditions.   However, it is possible to devise calculation procedures in which the hyperbolic
flows can be treated in the same manner as elliptic
flows; hence we shall not treat the hyperbolic flows
as a separate class.)   When the fluid velocity is
predominantly in one direction, the flow can be
treated as parabolic.   Then it is possible to march
in that direction from an upstream station to a downstream one predicting quantities plane by plane.
Such flows will be dealt with in Part II of this
article.   In Part I, we shall be concerned with
flows that do not necessarily possess this quality.

   We do not particularly distinguish between
steady and unsteady flows.   If the calculation procedure can handle unsteady flows, a steady-state
solution can be obtained by merely continuing the
process till quantities cease to change with time.
Alternatively, the time step in the finite-difference
procedure can be made very large, which has the effect
of eliminating the time-dependence terms.

   Our field of inquiry is that of unsteady, three-
dimensional, elliptic flows.   By this choice, no
restrictions are made regarding the time dependence,
the dimensionality and the presence or absence of recirculation in a given direction.   Consequently all
flows of engineering interest can be accommodated in
this field.

   Once the calculation scheme has been devised,
its usefulness depends on how well the equations describe the physical phenomena important to a given
situation.   Some of the phenomena of interest and
their interactions are: the effect of turbulence on
momentum transfer and on the transfer of a scalar

quantity; the mechanism of a chemical reaction influenced by chemical kinetics and turbulence; influences of chemical reaction and body forces on turbulence; effect of temperature and concentration fluctuations on the rate of a chemical reaction; influence of radiation heat transfer on the enthalpy distribution; formation and growth or reduction of
solid or liquid particles in a gas flow; their
effect on turbulence and other phenomena; behaviour
of non-Newtonian fluids. Satisfactory mathematical
models for many of these phenomena do not exist at
present; but current research suggests that these
deficiencies may be removed in a few years' time.

2.2   *Available Calculation Procedures*

Most attention in this article is given to the
calculation procedure developed by the author and his
colleagues. First, however, let us consider what
other procedures are available in the literature.

An obvious characteristic that can be used to
classify the calculation methods is the choice of the
dependent variables which describe the flow field.
For two-dimensional flows, stream function and vorticity are quite frequently used as the dependent
variables in preference to the two velocity components
and the pressure. For three-dimensional flow situations, stream function does not exist. A related
method employs the three components of vorticity and
the three components of a vector velocity potential
as the dependent variables. Such a procedure is
described by Aziz and Hellums [1], and is also
adopted by Mallinson and de Vahl Davis [2]. Although
for two dimensions the vorticity approach reduces the

number of variables from three to two, it seems to
increase this number for three dimensions. Moreover,
the absence of the pressure from the equations is inconvenient particularly for compressible flows.   Pre-
sumably for these reasons, all other calculation procedures employ the velocity components and the pressure as the dependent variables.

      In the velocity-pressure system, the distinction
between two and three dimensions is very little, and
hence methods primarily published for and applied to
two-dimensional flows can easily be extended to
three-dimensional flows.

      The differences between various velocity-pressure
methods lie in the formulation of the finite-
difference equations and in their subsequent solution.
In some procedures, all the dependent variables are
calculated for the same grid points; in others, the
locations for the calculation of the velocity components are displaced (or "staggered") relative to
the locations for which pressure is calculated.   For
time-dependent problems, some methods use explicit
schemes which limit the permissible size of time
step, while others employ implicit schemes which are
free from the time-step limitation.   Many different
practices are used for the solution of the finite-
difference equations.

      Most procedures of the velocity-pressure variety
involve the solution of the momentum equations (to
obtain the velocity components) and of the "Poisson
equation for pressure".   The latter is sometimes
solved by a direct method such as the Fourier method.

      The available velocity-pressure methods include
the procedures by: Harlow and Welch [3], Thommen [4],

Chorin [5], Williams [6], Deardorff [7], and Amsden
and Harlow [8].   It is not intended here to describe
and compare the details of these procedures.   The
method to be described in Sec. 3 embodies the best
features of the available methods and contains several novelties.

The methods mentioned above have all been applied
only to laminar flow, except that Deardorff [7] obtained the velocity fluctuations in a turbulent flow
by actually calculating the unsteady laminar motion
of fluid elements.   Application of these methods in
conjunction with turbulence models or chemical reaction in three-dimensional situations seems to be
rare;   only Zuber [9] has adapted the procedure of
Harlow and Welch [3] to a three-dimensional furnace.

The calculation method that is made the central
theme of this article is described in: Patankar and
Spalding [10], Caretto *et al.* [11], and Patankar and
Spalding [12].

3.   THE CALCULATION PROCEDURE

3.1   *The Equations to be Solved*

Prediction of the flow situations described in
Sec. 2.1 involves the solution of a set of equations.
Some of these equations are differential;   others are
algebraic.   The differential equations usually
result from conservation principles such as conservation of mass, Newton's second law of motion, the
first law of thermodynamics, conservation of a chemical species, and so on.   The diffusion fluxes in
these equations are given by additional laws (which

are generally differential) such as the Stokes viscosity law, the Fourier law of conduction and Fick's
law of diffusion.   For turbulent flow, the governing
differential equations are derived by "time-averaging"
the basic conservation equations for unsteady laminar
flow.   The resulting equations have, as their dependent variables, the time-mean values of the flow
properties (such as velocity, enthalpy, concentration), or the correlations of various fluctuating
quantities (such as the kinetic energy of the fluctuating motion, components of the Reynolds stress
tensor, the mean-square fluctuations of concentration).

The algebraic equations usually provide the link
between the dependent variables of the differential
equations and other quantities in these equations
such as fluid properties, exchange coefficients and
sources.   These equations are obtained from thermo-
dynamic relationships like the equation of state,
from hypotheses about the physical processes, or from
generalizations of experimental data.

How many differential equations should be solved
for a given flow situation is decided by the auxiliary relationships that it is reasonable to employ for
that problem.   If the flow is laminar, and if the
density and viscosity are uniform, then only the
equations of conservation of mass and momentum need
to be solved.   If the density is non-uniform and
depends on enthalpy and concentration, it is necessary to solve additional differential equations for
these quantities.   For turbulent flow, the (time-
mean) momentum equations contain the Reynolds
stresses;   the hypothesis for obtaining these stresses

may involve the solution of further differential equations.

Almost all the differential equations that are relevant here can be cast into a general form which we write as:

$$\frac{\partial(\rho\phi)}{\partial t} + \text{div}\,(\vec{G}\,\phi) = \text{div}\,(\Gamma_\phi\,\text{grad}\,\phi) + S_\phi \;;\quad (3.1)$$

here ϕ is the general dependent variable, $\vec{G}$ is the mass-flux vector, Γ_ϕ is the "exchange coefficient" for the diffusion of ϕ, and S_ϕ is the source term.* The expressions for Γ_ϕ and S_ϕ depend on the physical meaning given to ϕ and on the contents of its governing equation. It can be seen that the units of Γ_ϕ are those of the viscosity; in fact, Γ_ϕ is often obtained by dividing the viscosity by the appropriate Prandtl or Schmidt number. It should be noted that the use of Eq. (3.1) does not restrict us to phenomena which are governed by a gradient-type diffusion law; for, the part of the diffusion flux that does not obey the gradient law can be expressed as a part of S_ϕ. Indeed, the expressions for S_ϕ often contain terms that originate from a diffusion process.

In turbulent flow, the differential equation for a time-mean property ϕ contains additional terms which are often considered to represent the diffusion due to turbulent mixing; if this diffusion is expressed by a gradient law, Γ_ϕ can be thought of as the "effective" exchange coefficient, which takes

---

* A complete list of mathematical symbols and their definitions is provided under Nomenclature on p. 75.

account of the turbulent as well as the molecular
diffusion process.

When ϕ stands for the velocity component in a
given direction, Eq. (3.1) becomes a momentum
equation.   Since we shall need to pay special attention to the momentum equations, we separately write
the momentum equation for direction   i   as:

$$\frac{\partial(\rho V_i)}{\partial t} + \mathrm{div}(\vec{G}V_i) = \mathrm{div}(\Gamma_v \mathrm{grad}\ V_i) - \vec{i}_i \cdot \mathrm{grad}\ p + S_i\ ;$$

$$(3.2)$$

here V_i is the velocity component in direction   i,
$\vec{i}_i$ is the unit vector in that direction, and   p   is
the pressure.   It can be seen that Γ_v is the viscosity;   it should also be noted that the effect of
the viscous stresses on the direction-i momentum is
not fully expressed by $\mathrm{div}(\Gamma_v \mathrm{grad}\ V_i)$ and hence S_i
should in general contain terms arising from viscous
stresses.

The difference between equations (3.1) and (3.2)
is that the latter contains a pressure-gradient term.
The pressure   p   is one of the variables we need to
solve for, but the way to the solution appears to be
rather indirect at this stage.   The correct pressure
distribution is that which would result in a velocity
field (from the momentum equations) and a density
field (from the equation of state) that will satisfy
the equation for conservation of mass, which is:

$$\frac{\partial \rho}{\partial t} + \mathrm{div}\ \vec{G} = 0\ .$$

$$(3.3)$$

The device we employ to convert this indirect speci-

fication into a more direct solution algorithm is
described in Sec. 3.5.

   The source term S_i in the momentum equation
(3.2) can consist of, in addition to the viscous-
stress terms mentioned above, various body forces
and distributed resistances.   The distributed-
resistance concept is useful for calculating flows in
packed beds, in shell-and-tube heat exchangers, in
the subchannels of nuclear fuel-rod bundles, and in
similar situations where numerous solid objects offer
a resistance to fluid flow.

3.2   *The Basic Finite-difference Scheme*

   Here we describe how the general differential
equation for the variable ϕ,   Eq. (3.1), can be cast
into a finite-difference form.   First a grid is constructed which fills the domain of interest;   our aim
is to calculate the values of ϕ at the nodes of the
grid.   A rectangular grid is used such that the grid
lines are parallel to the three Cartesian co-ordinate
axes.   Other choices of grids are possible.   Figure
1 shows a part of the three-dimensional grid.   The

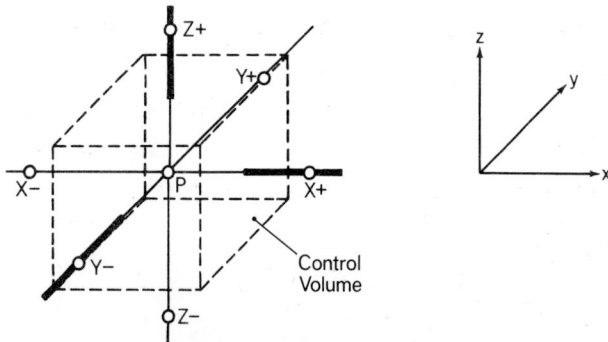

Fig. 1   The grid and a control volume.

grid point P has the points X+, X-, Y+, Y-, Z+
and Z- as its neighbours.   It may be noted that
the distances such as PX+ and PX- are not necessarily equal.   Also shown in Fig. 1 is a control volume surrounding the point P.   The control-volume
faces pass midway between P and the corresponding
neighbours.   Our finite-difference equation is
formed by integrating Eq. (3.1) over this control
volume.

The grid structure shows the discretization of
space.   The time is discretized as well; from the
known values of ϕ's at time t, we wish to obtain
the values at the next instant t + Δt.   When only
the steady-state solution is of interest, Δt may be
made very large.

We shall call the two terms on the left-hand
side of Eq. (3.1) the unsteady term and the convection term respectively; the terms on the right-hand
side will be referred to as the diffusion term and
the source term.   In the integration of Eq. (3.1)
over the control volume, the unsteady term will contribute to a volume integral and the convection and
diffusion terms will give rise to surface integrals
over the faces of the control volume; the source
term may produce volume and surface integrals depending upon the actual expression for S_ϕ.   To express
these integrals in terms of the values of ϕ at the
grid points, we need to assume the variation of ϕ
between the grid points.   In choosing this variation
we should ensure the compatibility of surface integrals between adjacent control volumes; for example,
the expression for the flux across the face between
the grid points P and X+ in Fig. 1 should be the

same irrespective of whether the control volume surrounding P or the one around X+ is being considered.   Within the limitations of this rule, our
policy is to choose the simplest variation of ϕ so
that the resulting difference equations have minimum
complication.

The unsteady term.   While integrating the term
$\partial(\rho\phi)/\partial t$ over the control volume, we assume that the
values of ϕ and ρ at the central point P prevail over the entire control volume.   Then the
finite-difference representation of the unsteady term
becomes:

$$(\rho_P \ \phi_P - \rho_P^o \ \phi_P^o)(\Delta x \ \Delta y \ \Delta z) \ / \ \Delta t \ , \qquad (3.4)$$

where $(\Delta x \ \Delta y \ \Delta z)$ is the volume, and the superscript
o denotes the values at time t, while the absence
of this superscript implies the values at t + Δt.

The convection and diffusion terms.   We consider
the terms div $(\vec{G} \ \phi)$ and div $(\Gamma_\phi \ \text{grad} \ \phi)$ together,
because they give rise to surface integrals, for
which the compatibility requirements should be obeyed.
Further, as discussed in Sec. 3.3 below, a proper
representation of these terms is essential to the convergence of the numerical procedure.

Our basic formulation of the convection and diffusion terms can be explained by considering the
transport across one face of the control volume.
Figure 2 shows the face of area A_x normal to the x
direction; this lies midway between the grid points
P and X+, which are a distance δx+ apart.   We
assume that ϕ varies *linearly* between P and X+.
Then the contribution C_{x+} by this face to the inte-

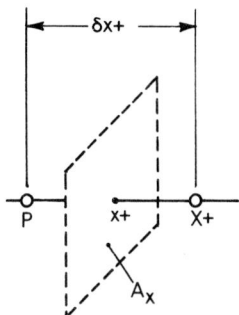

Fig. 2   The face of the control volume across which the x-direction convection and diffusion fluxes are considered.

gral of   div $(\vec{G}\phi - \Gamma_\phi \text{ grad } \phi)$ over the control volume is given by:

$$C_{x+} = (\rho u)_{x+} \, A_x (\phi_p + \phi_{X+})/2 - (\Gamma_\phi)_{x+} \, A_x (\phi_{X+} - \phi_p)/\delta x+ \; ;$$

$$(3.5)$$

here u is the x-direction velocity, so (ρu) is simply the component of $\vec{G}$ in the x direction. It will be shown in Sec. 3.3 that we need to modify Eq. (3.5) under certain circumstances.

The values of ϕ appearing in Eq. (3.5) are to be regarded as the values at time $t + \Delta t$; thus we use a "fully-implicit" formulation.

The source term.   The finite-difference representation of the source term depends on the actual expression for S_ϕ.   When the expression is known, its integral over the control volume can be worked out in a straightforward manner.

The complete difference equation.   When the finite-difference representation of all the terms in Eq. (3.1) has been worked out we can obtain the general difference equation of the following form:

$$a_p\phi_p = a_{X+}\phi_{X+} + a_{X-}\phi_{X-} + a_{Y+}\phi_{Y+} + a_{Y-}\phi_{Y-} + a_{Z+}\phi_{Z+} + a_{Z-}\phi_{Z-} + b \ ,$$

$$(3.6)$$

where the ϕ values correspond to time $t + \Delta t$; the value of ϕ_p at time t has been absorbed in b. Although Eq. (3.6) has the appearance of a linear equation, it must be remembered that the coefficients themselves depend on the values of different ϕ's. This non-linearity will be handled by an iteration scheme, in which the coefficients are recalculated in every iteration cycle.

3.3 The "Hybrid" Formulation for the Combined Effect of the Convection and Diffusion Terms

Equation (3.6) shows that the value of ϕ at the central point P is related to the values of ϕ at the six neighbouring points. It we assume, for convenience, that the source term does not depend on the neighbour-point ϕ values, then the neighbour coefficients a_{X+}, a_{X-}, a_{Y+}, a_{Y-}, a_{Z+} and a_{Z-} must represent the processes of convection and diffusion only. Indeed, we can then deduce from Eq. (3.5) that the coefficient a_{X+} in Eq. (3.6) is given by:

$$a_{X+} = (\Gamma_\phi)_{X+} A_X \Big/ \delta x+ - (\rho u)_{X+} \Big/ 2 \ . \qquad (3.7)$$

It can be seen that the first term on the right-hand side will always be positive, but the second term can be positive or negative. Thus, if u is positive and large, it is possible that a_{X+} will be negative. This means that an *increase* in the neighbour value ϕ_{X+} will lead to a *decrease* in ϕ_p. This is obviously a physically-unrealistic result.

The cause of this implausible result can be traced to our assumption of a *linear* variation of ϕ between P and X+. This assumption is quite reasonable when the convection influence is small, but becomes incorrect for large convection rates. Large rates of mass flow tend to establish the "up-wind" value of ϕ in the downwind region. Thus, when u is large and positive, the value of ϕ convected across the face in Fig. 2 is *not* $(\phi_P + \phi_{X+})/2$ as our linear variation implies, but it is nearly ϕ_P itself. For large negative u, the value ϕ_{X+} will be convected across the face. Further, under these conditions the diffusion across the face will be negligible.

Having decided what would be a reasonable representation of convection and diffusion at very small and very large convection rates, we can proceed to formulate a scheme for the entire range. We use a "hybrid scheme" which is a combination of the so-called central- and upwind-difference schemes. Its rationale is explained by Spalding [13], and it has been widely tested in all the calculation procedures with which the author is associated.

The hybrid formulation replaces Eq. (3.5) by a three-part formula which expresses the convection-and-diffusion flux C_{x+} across the face between P and X+ by:

$$\left| (\rho u)_{x+} \right| \delta x+ \Big/ (\Gamma_\phi)_{x+} < 2:$$

$$C_{x+} = (\rho u)_{x+} A_x (\phi_P + \phi_{X+}) \Big/ 2 - (\Gamma_\phi)_{x+} A_x (\phi_{X+} - \phi_P) \Big/ \delta x+ \quad ;$$

$$(3.8)$$

$(\rho u)_{x+}\ \delta x + / (\Gamma_\phi)_{x+}\ \geqslant 2:$ $C_{x+}\ =\ (\rho u)_{x+}A_x\phi_P\ ;$

$(\rho u)_{x+}\ \delta x + / (\Gamma_\phi)_{x+}\ \leqslant -2:$ $C_{x+}\ =\ (\rho u)_{x+}A_x\phi_{X+}.$

$$(3.8)$$
$$\text{cont.}$$

It is easy to work out that as a consequence of Eq. (3.8) the expression for the coefficient a_{X+} will become:

$$\left|(\rho u)_{x+}\right|\ \delta x + / (\Gamma_\phi)_{x+}\ <\ 2:$$

$$a_{X+}\ =\ (\Gamma_\phi)_{x+}A_x/\delta x + -\ (\rho u)_{x+}A_x\ /\ 2 \qquad\qquad ;$$

$$(\rho u)_{x+}\ \delta x +\ /\ (\Gamma_\phi)_{x+}\ \geqslant 2: \qquad a_{X+}\ =\ 0 \qquad\qquad ;$$

$$(3.9)$$

$$(\rho u)_{x+}\ \delta x +\ /\ (\Gamma_\phi)_{x+}\ \leqslant -2: \qquad a_{X+}\ =\ -(\rho u)_{x+}A_x\ .$$

Although the need for the modification of our differ- ence formulae is introduced here on the basis of physical realism, it can be shown that such a modifi- cation is essential also to the convergence of the iteration scheme used to solve the difference equations.

There are successful procedures which do not in- clude a modification of the kind described in this sub-section.   These procedures employ ways of ensur- ing that one always operates on the first part of the above three-part formula.   If one uses a sufficiently fine grid (i.e. $\delta x+$ is small enough), then the con- dition for the first part can be satisfied; but this practice often proves to be prohibitively expensive of computer time and storage.   Another practice is to augment Γ_ϕ artificially (the "artificial- viscosity" method), which again keeps one on the first part of Eq. (3.8) or (3.9); but then one has to know beforehand by how much the value of Γ_ϕ should be augmented, and this augmentation is carried to those flow regions where it is not needed.   Thus, the hybrid scheme seems to be preferable to the alternative practices mentioned above.

3.4   *Solution of the Difference Equations*

When we have constructed algebraic equations like Eq. (3.6) for all the nodes in the calculation domain, then the next task is to solve this set of equations.   There are various methods for performing this task, some more suitable than others.   Direct solution by the Gauss elimination method requires too large computer storage and time.   The Gauss-Seidel method of successive substitution converges rather slowly especially when the number of equations is large.   We therefore employ a line-by-line solution procedure, which is outlined below.

Consider the equations for the nodes along one grid line.   Let us assume that the values of ϕ for these nodes are unknowns, but that those for the nodes along the four neighbouring lines are known. Then it is easy to see that we can solve for the unknown ϕ's by the use of the well-known tri-diagonal-matrix algorithm (TDMA).   For the values of ϕ along the neighbouring lines (which are considered to be temporarily known), we simply use the best estimates, which are often the values from the previous iteration.   In this manner, we traverse along all lines in one direction, then (using the resulting solution as our best estimate) repeat the process along the second direction, and finally perform the same operation along the third direction.

It should be remembered that we do not seek a
very accurate solution of the algebraic equations in
a given iteration; because the coefficients in these
equations are only tentative and need to be recalculated as a result of the change in the values of ϕ.
Usually one or two traverses in each direction are
found to be sufficient before the coefficients are
recalculated.

The line-by-line procedure is particularly
effective when the magnitudes of the coefficients in
different directions are largely unequal; in such a
case, the traverse in the direction of the largest
coefficients yields an almost exact solution of the
algebraic equations.

3.5   *Handling the Pressure Linkage*

As mentioned before, a special difficulty arises
from the fact that the momentum equations contain the
pressure-gradient term and that the pressure does not
appear explicitly in the continuity equation.   Our
procedure for handling this pressure linkage has been
given the name SIMPLE (Semi-Implicit Method for
Pressure-Linked Equations).

In Fig. 1 we showed a portion of the grid for
the nodes of which we calculate the values of ϕ.
The three velocity components however are calculated
for points which lie midway between these nodes.
These points are shown in Fig. 3 as x+, x-, y+, y-,
z+ and z- and are denoted by small arrows.   The
pressure is calculated for the node points, which are
shown as P, X+, X-, etc.. It will be seen that
the x-direction velocity u is calculated for the
midpoint of the x-direction link joining two nodes.

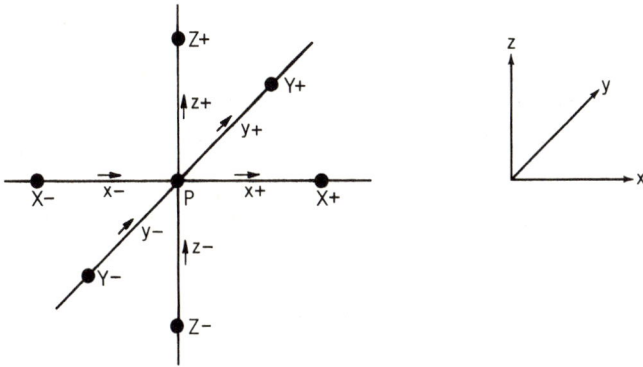

Fig. 3   The locations for which the velocity
              components are calculated.

The advantage of these "staggered" locations for the
storage of different variables is two-fold: firstly,
for the control volume in Fig. 1, the velocity component normal to a face is calculated for a point on
that face, which makes it convenient to write the
mass conservation for that control volume;   secondly,
in the momentum equations, the difference of pressure
between two *adjacent* nodes can be used to "drive" the
velocity component for the point midway between them.
The finite-difference equations for the velocity components u, v and w are obtained in the same
manner as Eq. (3.6), but there are two different features.   Firstly, the control volumes are centered
around the points shown by arrows in Fig. 3;   and
secondly, the pressure-gradient term is processed
separately.   Thus, the difference equation for the
velocity u has the form:

$$a_{x-} u_{x-} = a_{x+} u_{x+} + \ldots\ldots + b + d_{x-} (p_{X-} - p_p), \quad (3.10)$$

where the missing terms denote the influence of the
remaining five neighbours of u_{x-} (which are not
shown in Fig. 3), and d_{x-} is simply the area on
which the pressure difference acts.

Since the values of the pressure p appearing
in Eq. (3.10) are not known beforehand, it is our
practice to substitute our best estimate of the
pressure field, denoted by p*, into Eq. (3.10) and
obtain a preliminary velocity field u*, v* and w*.
Thus, we have:

$$a_{x-}\, u^*_{x-} = a_{x+}\, u^*_{x+} + \; \ldots \; + b + d_{x-}\, (p^*_{X-} - p^*_P). \qquad (3.11)$$

The true pressure p is related to the estimated
pressure by:

$$p = p^* + p' \qquad , \qquad (3.12)$$

where p' is the so-called pressure correction. We
further suppose that the true velocity u is con-
nected to u* via:

$$u_{x-} = u^*_{x-} + D_{x-}\, (p'_{X-} - p'_P) \qquad , \qquad (3.13)$$

where D_{x-} is simply d_{x-}/a_{x-}. Similar equations
hold for v* and w*. Now we turn to the question
of obtaining p'.

We derive a finite-difference equation for p'
by the use of the continuity equation. Although it
is possible in this derivation to account for a
pressure-dependent density (and this has been done by
the author elsewhere [14]), we shall assume here, for
convenience, that the density does not depend on the

local pressure and can be calculated from the values
of other variables.

Integration of the continuity equation (Eq.
(3.3)) over the control volume around the point P
gives us the following finite-difference equation:

$$(\rho_P - \rho_P^o)(\Delta x \; \Delta y \; \Delta z)/\Delta t = \{(\rho u)_{x-} - (\rho u)_{x+}\} \; A_x$$
$$+ \{(\rho v)_{y-} - (\rho v)_{y+}\} \; A_y$$
$$+ \{(\rho w)_{z-} - (\rho w)_{z+}\} \; A_z \quad , \quad (3.14)$$

where ρ_P and ρ_P^o represent the values of density
at times $t + \Delta t$ and t respectively.

Now, if we substitute from equations like (3.13)
for the u's, the v's and the w's in Eq. (3.14),
we get the difference equation for p', which has
the form:

$$a_P p_P' = a_{X+} p_{X+}' + a_{X-} p_{X-}' + a_{Y+} p_{Y+}' + a_{Y-} p_{Y-}'$$
$$+ a_{Z+} p_{Z+}' + a_{Z-} p_{Z-}' + b \; . \quad (3.15)$$

3.6   *Summary of the Solution Procedure*

Now we summarize the various steps in the whole
calculation procedure.   We select the dependent
variables, the differential equations and the auxiliary relations.   Then we are in a position to form
the required finite-difference equations.   Their
solution proceeds by the cyclic repetition of the
following steps:

(i)   Provide initial estimates of the values of
         all the variables including pressure p*.

Hence calculate auxiliary variables such as density, viscosity etc..

(ii)   Solve the momentum equations like Eq. (3.11) to obtain u*, v*, w*.

(iii)   Solve the pressure-correction equation (Eq. (3.15)) to obtain p'.

(iv)   Calculate the pressure p from Eq. (3.12) and the corrected velocities u, v, w from equations like (3.13).

(v)   Solve the equations of the form (3.6) for all other dependent variables.

(vi)   Regard the new values of the variables as improved estimates and return to step (i). Repeat until convergence.

3.7   Miscellaneous Matters

Underrelaxation.   Our finite-difference equations, when considered linear, are so constructed as to guarantee convergence of the line-by-line (or, the point-by-point for that matter) solution procedure.   However, divergence can result from the changes in the coefficients in the difference equations from one iteration cycle to the next.   To combat any such tendency, it is often necessary to employ some form of underrelaxation.   This is especially useful when the time step chosen is rather large.   Indeed, the use of a moderate time step, even when only the steady-state solution is required, is one method of achieving underrelaxation.   All methods of underrelaxation try to reduce the change in the value of a variable during one iteration.   In addition to the dependent variables, auxiliary quantities like the density and the viscosity can be

underrelaxed with advantage.   Overrelaxation, which
is the counterpart of underrelaxation, is sometimes
used to speed up convergence;   in most problems of
our interest however, the interlinkages between various equations are so strong that it is usually
necessary to slow down the changes rather than to en-
courage them.

   Boundary conditions.   The finite-difference
equations for a given variable form a solvable set
only after the information about the boundary conditions is inserted into them.   If the values of the
dependent variable are known for the boundary nodes,
all that we need to do is to use these values where
they appear as the neighbours of the internal nodes
near the boundary.   When the boundary condition is
in the form of a specified diffusion flux or some
other relationship, a different practice is needed.
Essentially we then form an additional algebraic
equation for each boundary unknown and solve these
equations simultaneously with the regular difference
equations.   Alternatively, one can first modify the
difference equations for the near-boundary nodes by
the use of the boundary-condition information and
thus eliminate the unknown values;   then the set of
finite-difference equations can be solved.

   The boundary conditions for the pressure-
correction equation require special mention.   Usually
either the velocity or the pressure is specified at a
boundary.   When the velocity normal to the boundary
is specified, there would be no correction to this
velocity, and thus a boundary condition of zero normal gradient is appropriate for the   p'   equation.
When the pressure at the boundary is specified, we

can make the estimated pressure p* at the boundary
the same as the given pressure p;   then p' for the
boundary points must be zero, which is a convenient
boundary condition for the p' equation.

     Irregular boundaries.   Since we have used a
Cartesian co-ordinate system, it is most convenient
if the calculation domain has the shape of a rectangular box.   Yet many practical situations require
calculation domains with inclined or curved boundaries;   in addition, there may be obstacles in the
middle of the flow region.

     Some of the geometries can be handled by the use
of a suitable curvilinear orthogonal co-ordinate system;   and the required details of the finite-
difference procedure can be worked out in an analogous manner.

     It is however often more convenient to employ a
Cartesian co-ordinate system but to modify the difference equations wherever the boundary intersects
the grid in an irregular fashion.   This modification
takes the form in a computer program, of "overwriting"
the regular finite-difference coefficients by the
special ones for the near-boundary points.   Once the
coefficients are modified, the solution of the algebraic equations can proceed in the usual manner.

4.   MATHEMATICAL MODELS FOR PHYSICAL PHENOMENA

     The subject of mathematical models for various
physical processes of our interest is very vast.   It
is beyond the scope of this article to give a comprehensive account of the present-day state of the art.

We shall therefore make reference to other papers
where such information is available, and describe
only those models in detail which are relevant to
the examples to be presented later in this article.
We shall give attention to the models of turbulence,
combustion and radiation.

4.1   *Turbulence*

Reviews of models of turbulence can be found in
the book by Launder and Spalding [15] or in Ref. [16]
by the same authors.

If the Reynolds stresses are expressed via the
eddy-viscosity concept, we have the following relation for Cartesian co-ordinates:

$$\overline{\rho u_i' u_j'} = -\mu_t \left(\frac{\partial u_i}{\partial x_j} + \frac{\partial u_j}{\partial x_i} \right) + \frac{2}{3} (\mu_t \text{ div } V + \rho k) \delta_{ij} \ , \qquad (4.1)$$

where   u'   denotes the velocity fluctuation, $\overline{\rho u_i' u_j'}$
is one of the Reynolds stresses, μ_t is the turbulent viscosity, and   k   is the kinetic energy of
turbulence.   The subscripts   i   and   j   can take the
values   1,   2   or   3; u_1, u_2, u_3 are supposed to
be the time-mean velocities in the three co-ordinate
directions x_1, x_2 and x_3 respectively; δ_{ij} is
the Kronecker delta.

If we accept Eq. (4.1) as the basis of our calculation, then a formula is needed for the calculation of the turbulent viscosity μ_t.   This will be
provided in two ways below.

The mixing-length formula.   Usually, the
mixing-length hypothesis is used in circumstances
where only one velocity component is predominant and

where its variation in only one direction is appreci-
able.   We can however generalize the usual mixing-
length formula in the following manner:

$$\mu_t = \rho L^2 \left\{ \left(\frac{\partial u_i}{\partial x_j} + \frac{\partial u_j}{\partial x_i} \right) \frac{\partial u_i}{\partial x_j} \right\}^{\frac{1}{2}} , \qquad (4.2)$$

where   L   is the mixing length, and the tensorial
summation convention should be used inside the curly
brackets.

The variation of   L   is usually specified via an
algebraic formula.   Near a single wall,   L   is taken
as proportional to the distance from the wall.   When
the calculation domain is bounded by a number of
walls, account must be taken of the distances from
all these walls.   Buleev [17] has worked out formulae for the mixing length in enclosures formed by
many walls.   It is through the need for a prescrip-
tion of   L   that the shortcoming of the mixing-length
theory becomes clear.   Except for some very simple
geometries, it is very difficult to prescribe a
variation of   L   that will be in accordance with experimental data.   As a result, the utility of the
mixing-length hypothesis is limited to very simple
flow configurations.

The dependence of μ_t on the gradients of mean-
velocity components also restricts the applicability
of the mixing-length formula.   For, in the regions
of small velocity gradients, the formula predicts too
low values of the turbulent viscosity.   This becomes
particularly serious when the turbulent exchange coefficients for other quantities such as enthalpy and
concentration are taken to be proportional to the

turbulent viscosity.

    A two-equation model. A turbulent-viscosity formula that is not tied up with the mean velocity is:

$$\mu_t = \rho k^{\frac{1}{2}} \ell \qquad , \qquad (4.3)$$

where k is the kinetic energy of the fluctuating motion and ℓ is a length scale similar to the mixing length. The quantity k is to be obtained from a differential equation that describes the convection, diffusion, generation and dissipation of k. The length scale ℓ can be specified algebraically; but then the resulting model has the same deficiency as the mixing-length model: it becomes restricted to very simple geometries. For general three-dimensional flows, we solve an additional differential equation from which ℓ can be derived. The dependent variable of this equation is ε which is related to ℓ by:

$$\varepsilon = c_D \, k^{3/2} \Big/ \ell \qquad (4.4)$$

where c_D is a constant. Because this turbulence model now requires the solution of two differential equations, those for k and ε, we refer to it as a two-equation model.

    The form of these equations is that of Eq. (3.1). Therefore, instead of writing the full differential equations for k and ε, we give below the expressions for Γ_ϕ and S_ϕ for these equations.

ϕ	Γ_ϕ	S_ϕ
k	μ_t / σ_k	$g - \rho\varepsilon$
ε	$\mu_t / \sigma_\varepsilon$	$(c_1 g - c_2 \rho\varepsilon)(\varepsilon/k)$

Here σ_k and σ_ε can be considered as the turbulent Prandtl numbers for k and ε; they are usually taken as constants. c_1 and c_2 are two other constants. g is the rate of generation of k, and $\rho\varepsilon$ is the rate of dissipation. The expression for g is:

$$g = -\overline{\rho u_i' u_j'} \frac{\partial u_i}{\partial x_j} , \qquad (4.5)$$

where the summation convention should be used.

The values of the various constants in this model are adjusted with reference to experimental data. At present, the set that seems to give the best overall agreement with experimental data is:

$$c_D = 0.09, \quad c_1 = 1.44, \quad c_2 = 1.92,$$
$$\sigma_k = 1.0, \quad \sigma_\varepsilon = 1.3.$$

The turbulent exchange coefficient. Just as the Reynolds stresses appear in the time-averaged momentum equations, terms which may be regarded as representing the diffusion due to turbulent mixing appear in the time-averaged equations for other quantities such as enthalpy and concentration. When the

turbulent diffusion flux is expressed by a gradient
law, Γ_ϕ contains a contribution Γ_t, the turbulent
exchange coefficient. Usually Γ_t is expressed in
terms of μ_t by:

$$\Gamma_t = \mu_t / \sigma_t \qquad , \qquad (4.6)$$

where σ_t is the turbulent Prandtl or Schmidt number.
Normally, σ_t is given a uniform value of about 0.9.

More sophisticated models. Research into
models of turbulence has already gone beyond the two-
equation level. Models that require the solution of
differential equations for the six components of the
Reynolds stress tensor are being investigated by many
workers. Models which solve for the triple velocity
correlations or for three different length scales
have also been proposed. There are models which,
instead of using a turbulent Prandtl number, directly
solve for the turbulent diffusion flux of, say, enthalpy. Satisfactory methods have also been worked
out for the calculation of the mean-square fluctuations of temperature, concentration etc.. For a
description of these models, the reader should turn
to the references already mentioned.

4.2 Combustion

Prediction of a combustion process involves the
solution of the conservation equations for various
chemical species and for enthalpy, along with the
flow equations. The chemical-reaction schemes for
most common fuels, however, are very complex. The
difficulty is further enhanced by the lack of chemical-
kinetic data for the numerous reactions and by the un-

certainty about the influence of turbulence on the
rate of reactions.   A set of simplifying assumptions
about the combustion process constitute a model for
combustion.   The characteristics of the model that
we have employed in Sec. 5 will now be described.

   The participating chemical species.   We represent combustion by what is known as a "simple chemical reaction".   This involves only two reacting
species, namely fuel and oxygen, which combine in a
fixed proportion to produce a single species called
the product.   (In most practical systems, oxygen
comes from an air stream; nitrogen in that stream is
considered to be "product" because it is inert and
because its molecular weight is comparable with that
of the products of most hydrocarbon fuels.)

   The composite fuel-oxygen variable.   The mass
fractions of fuel and oxygen, m_{fu} and m_{ox}, are
governed by differential equations of the form (3.1).
The source S_ϕ represents the consumption of the
species by chemical reaction.   If a unit mass of fuel
requires i units of mass of oxygen for complete combustion, it is easy to see that, at any point in
space, S_ϕ for fuel will be (1/i) times S_ϕ for
oxygen.   Further, we can assume that Γ_ϕ is the same
for both m_{fu} and m_{ox}; this assumption is quite
reasonable for turbulent flow, and is often not incorrect for laminar flow.   With these considerations,
we can form a differential equation for the composite
quantity $(m_{fu} - m_{ox} / i)$, which will have the same
form as Eq. (3.1), but the S_ϕ for which will be
zero.   It is this dependent variable that we frequently use in combustion calculations.

The diffusion-flame assumption.   The solution
for the variable ($m_{fu} - m_{ox}$ / i) still does not enable us to obtain separate values of m_{fu} and m_{ox}.
For this, we should normally solve the equation for
m_{fu} (or m_{ox}) after having supplied the appropriate
expression for S_ϕ.   However, a short-cut is available in many practical situations.   If the rate of
reaction is assumed to be very large (which is the
case in many industrial furnaces), it is impossible
for both fuel and oxygen to co-exist at any point.
This means that at least one of the two mass fractions, m_{fu} and m_{ox}, must be zero at a point;
thus, if ($m_{fu} - m_{ox}$ / i) is positive, it must be
equal to m_{fu}, and if it is negative, it is simply
$-m_{ox}$ / i.   Flames governed by this assumption of a
very fast reaction are called diffusion flames.   The
example in Sec. 5.5 below will be of the diffusion-
flame variety.

Influence of chemical kinetics and turbulence.
Of course, not all combustion situations can be
handled by a diffusion-flame approximation.   In particular, when the supply stream is a mixture of fuel
and air, the existence of a flame or the possibility
of a blow-out is determined by the rate of reaction
which is influenced by a number of factors.   In such
cases, in addition to the equation for ($m_{fu}-m_{ox}$ / i),
we need to solve an equation for m_{fu}.   The expression for S_ϕ for m_{fu} is obtained from chemical
kinetics.   Turbulence also exerts an influence on
this reaction rate.   A preliminary model for this
influence was proposed by Spalding [18];   a refinement of the model and its use are reported by Mason
and Spalding [19].   This refinement includes the

effect of concentration fluctuations on the reaction
rate.   We shall however not present any calculations
which make use of such models.

4.3   *Radiation*

When the fluid has the ability to absorb, emit
and scatter thermal radiation, and when the temperatures are sufficiently high for the radiation to be
significant, the equation for the conservation of enthalpy should include a source term representing the
net gain of energy by radiation.   There are various
ways to calculate this term, and they employ different degrees of approximation.   These ways are discussed by Patankar and Spalding [20].   The method
we adopt is the so-called "six-flux model", which
will be briefly described below.

Let   I   and   J   stand for the forward and backward radiation fluxes in the   x   direction;   let   K,
L   and   M,   N   be the corresponding fluxes in the
y   and   z   directions respectively.   The rate of
change of   I   and   J   with distance is given by:

$$\frac{dI}{dx} = aE - (a+s) I + (s/6)(I+J+K+L+M+N) \quad , \quad (4.7)$$

and

$$\frac{dJ}{dx} = -aE + (a+s) J - (s/6)(I+J+K+L+M+N) , \quad (4.8)$$

where   E   is the black-body emissive power,   a   is
the absorption coefficient per unit length and   s   is
the scattering per unit length.   Combining these
equations, we get:

$$\frac{d}{dx} (I+J) = -(a+s)(I-J) \quad , \quad (4.9)$$

and

$$\frac{d}{dx}(I-J) = 2aE-(a+s)(I+J)+(s/3)(I+J+K+L+M+N). \quad (4.10)$$

Finally, we derive the following second-order differential equation for $(I+J)$:

$$\frac{d}{dx}\left\{\frac{1}{a+s}\frac{d}{dx}(I+J)\right\} = -2aE+(a+s)(I+J)$$

$$-(s/3)\left\{(I+J)+(K+L)+(M+N)\right\}. \quad (4.11)$$

On similar lines, we obtain equations for $(K+L)$ and $(M+N)$. It is straightforward to express these equations in finite-difference form and to solve them. The source term due to radiation for the enthalpy equation is then calculated from the values of E and $(I+J)$, $(K+L)$ and $(M+N)$.

5.   SOME PREDICTIONS OF THREE-DIMENSIONAL ELLIPTIC
     FLOWS

We now present the results of some computations performed by the use of the procedure described and the mathematical models outlined. These results are in accordance with our qualitative expectations, but their quantitative agreement with experimental data remains to be examined. Wherever a two-dimensional counterpart of the physical situation can be constructed, comparisons have been made with experimental data or with earlier solutions.

In this section, five different flow situations are considered. In the first two, the flow is lami-

nar.   The third and the fourth cases include a predominant effect of distributed resistance.   The
fourth case has some unsteady features also.   The
fifth problem is the most complex of all.   It includes turbulence, combustion and radiation.   It is
presented here to demonstrate that our calculation
procedure can already handle a large number of simultaneous influences.

     All the predictions are taken from recently-
published papers;   only the most important details
are given here, therefore.   For more information,
the reader should turn to the original papers.

5.1   *Laminar Flow in a Cavity*

     Here we consider a shallow rectangular cavity
filled with a fluid, across the free surface of which
moves a strip of solid having a width less than that
of the cavity.   The flow is considered laminar, and
the density and the viscosity are taken as uniform.

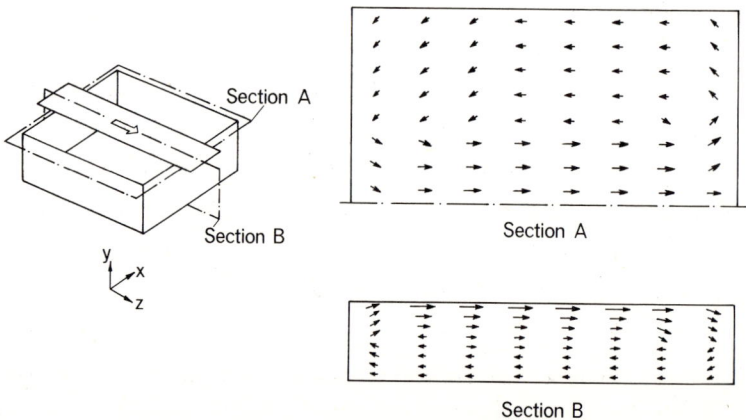

Fig. 4   Flow in a cavity created by the movement
           of a plate over the free surface.

Figure 4 shows, by velocity vectors, the flow patterns
on a horizontal and a vertical plane.   The Reynolds
number based on the strip velocity and the depth of
the cavity is equal to 100 for this calculation;
but it is worth remarking here that it has been possible to obtain convergent solutions for arbitrarily
high Reynolds numbers.   The computer time for this
problem with an $8 \times 8 \times 8$ grid is about 30 seconds
on a CDC 6600 computer.   Figure 4 is taken from Ref.
[20] and is one of the many cases presented in Ref.
[12].

5.2   *Flow of Wind Over a Building*

Figure 5 shows another laminar-flow situation.
Wind having a boundary-layer-type velocity profile
approaches a rectangular building.   Smoke is introduced at an upstream point.   The flow field and the
smoke concentration are calculated.   The results are
shown in Figs. 6 and 7.   The separation upstream and

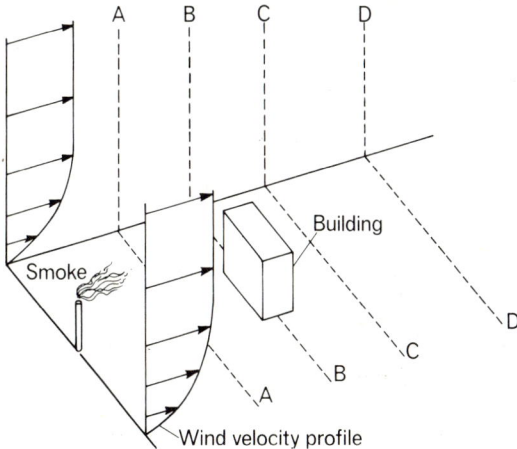

Fig. 5   Flow of wind over a building.

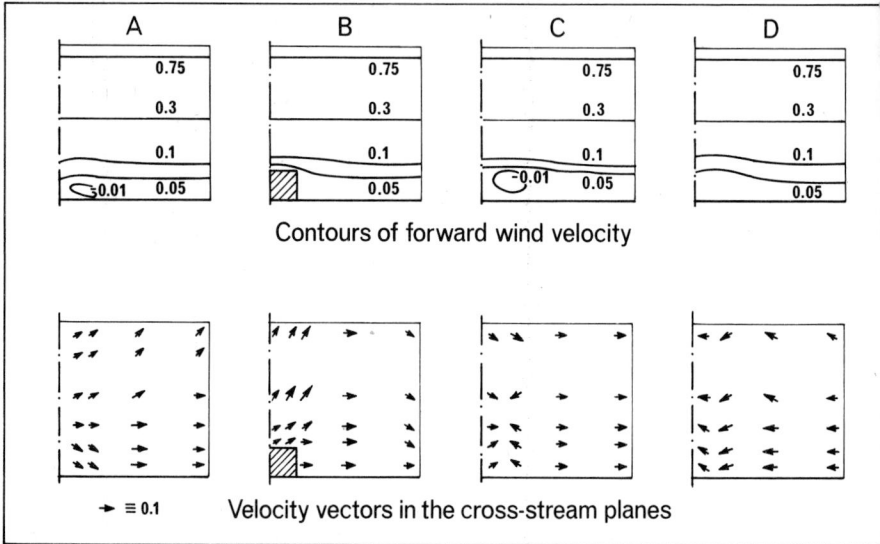

Fig. 6   Flow field around the building.

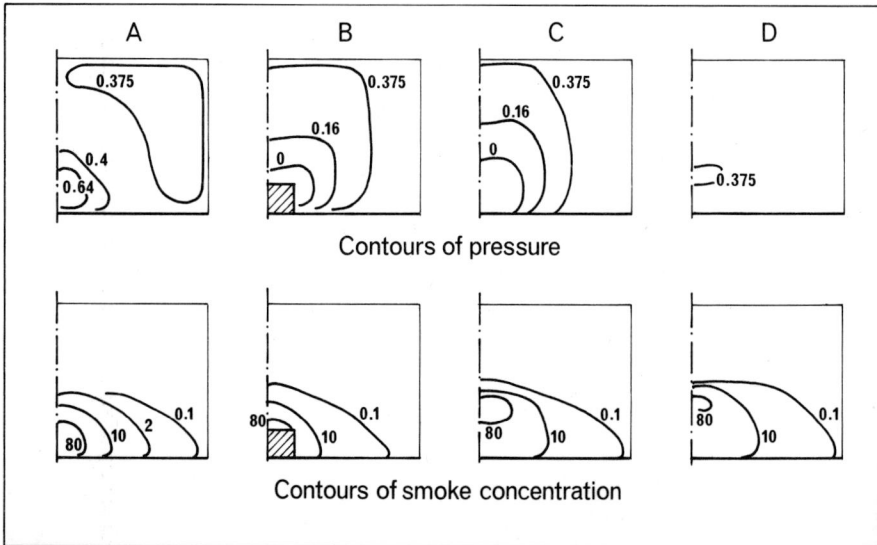

Fig. 7   Pressure and concentration fields around
              the building.

downstream of the building can be noted.   The computer time for this problem is similar to that quoted
in Sec. 5.1.   For more information the reader may
turn to Ref. [11].

5.3   Flow in a Packed Bed

Now we present an example where the flow is determined by the inertia and the distributed-
resistance terms.   It is the flow in a packed bed
shown in Fig. 8.   Fluid enters the bed through an
inclined slot in one wall and leaves through the open
area at the top.   The distributed resistances are
calculated on the assumption that the flow experi-
ences a constant friction factor in the interspaces
within the packing material.   The resulting velocity
field is shown in Fig. 8 by velocity vectors in two
vertical planes.   The recirculation in the neighbour-

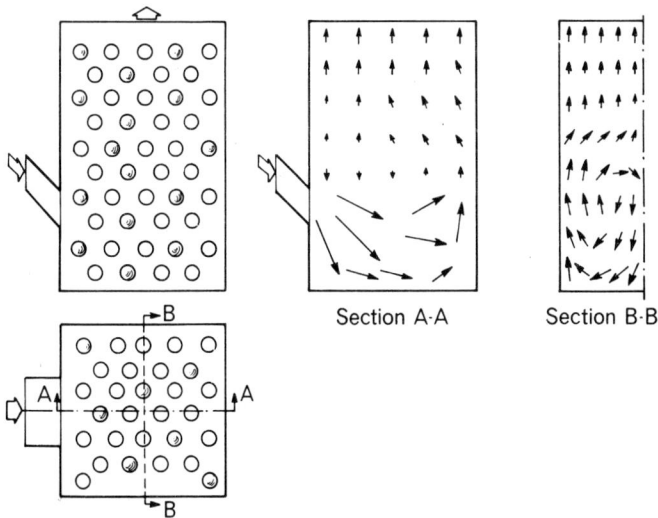

Section A·A   Section B·B

Fig. 8   Flow in a packed bed with inflow from an
              inclined tuyère.

hood of the injection slot can be clearly seen.   This
example is taken from Ref. [20].

5.4   Flow in a Shell-and-Tube Heat Exchanger

The distributed-resistance technique is applied
to a more complex situation by Patankar and Spalding
in Ref. [21].   Here we present some of their results.
Figure 9 shows the geometry of the shell-and-tube heat
exchanger for which the flow and the temperature
fields are calculated.   The tubes are fitted in a
five-pass arrangement.   One tube in Fig. 9 is really
a schematic representation for a large number of
small tubes that will exist in a real situation.   The
distributed resistance to the shell-side fluid is
greater in the x and y directions than in the z
direction.   We calculate the three velocity components and the pressure for the shell-side fluid; we
also calculate two temperature fields: the temperature of the shell fluid and the temperature of the

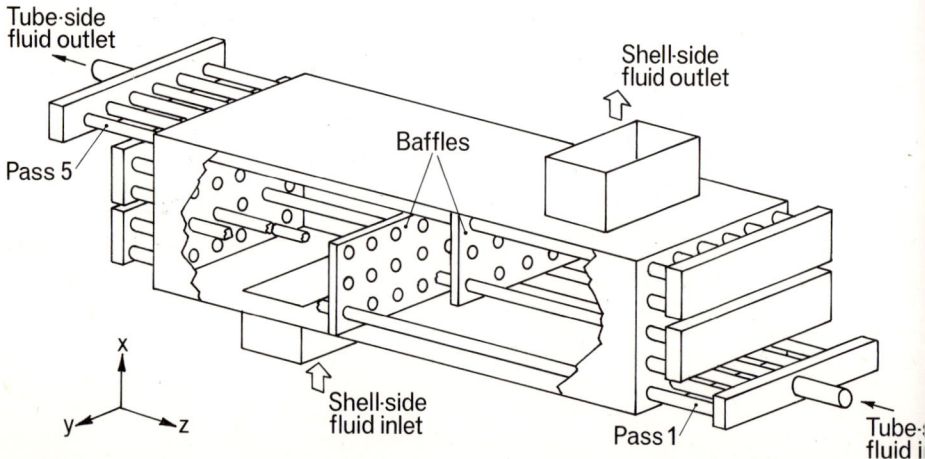

Fig. 9   The geometry of the heat exchanger.

tube fluid.   Heat transfer between the shell fluid
is calculated by assuming a uniform overall heat-
transfer coefficient.   Computations for both transient and steady-state behaviour are reported in Ref.
[21].   Here Figs. 10 and 11 display some of the
steady-state results, and Fig. 12 gives an idea of a
transient calculation.

The velocity field in the heat exchanger is
shown in Fig. 10 by means of velocity vectors in different planes.   The recirculations near the two ends
of the shell and the deflection of the flow by the
baffles can be clearly seen.

Results for the temperature distribution for a
particular value of the heat-transfer coefficient are

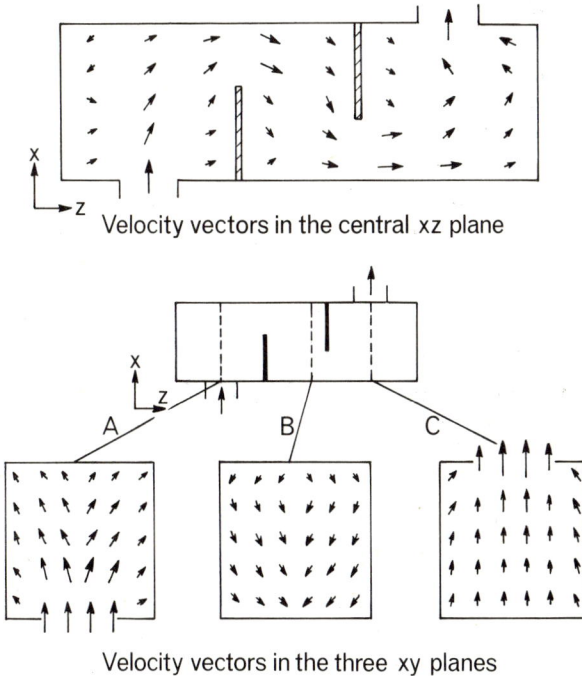

Velocity vectors in the central xz plane

Velocity vectors in the three xy planes

Fig. 10   The flow field in the heat exchanger.

(a)   Variation of the shell·fluid temperature

(b)   Variation of the tube·fluid temperature

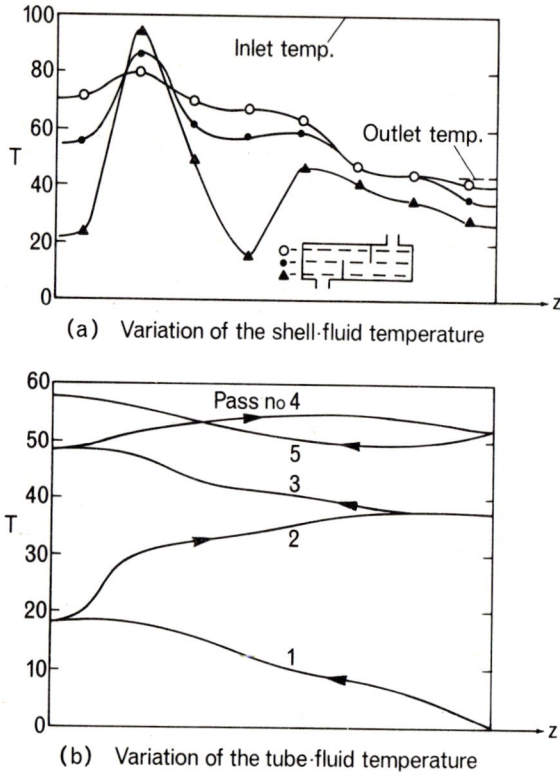

Fig. 11   Temperature distribution on the central
xz plane of the heat exchanger.

shown in Fig. 11.   The shell-fluid temperature is
plotted for three chosen lines on the central   xz
plane.   For the tube-fluid temperature, its variation along the five passes is shown for the central
xz   plane.   It can be seen that the shell-fluid temperature has a large spatial variation;   indeed,
there are locations where the shell-fluid temperature
drops below even the outlet temperature of the shell
fluid.   As a result of this, the tube-fluid temperature does not rise monotonically as we traverse along
the five passes.

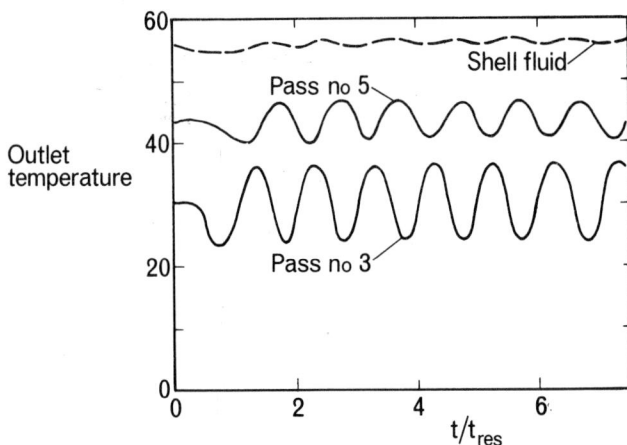

Fig. 12   Variation of outlet temperatures created by a periodic variation of the inlet temperature of the tube fluid.

Figure 12 is included here to give a taste of the unsteady-state behaviour. The inlet temperature of the tube fluid is varied in a periodic manner with an amplitude of 20 and a period of one residence time of the tube fluid. The variation with time of some of the outlet temperatures is plotted in Fig. 12. The decay of the amplitude is noticeable and so is the phase lag.

5.5   Flow Pattern and Radiation in a Gas-Turbine Combustion Chamber

Now we come to the last example in this section and to the most complex one as well. This is presented here mainly to demonstrate that it is already possible to make computations for a situation which is simultaneously influenced by the processes of turbulence, combustion and radiation.

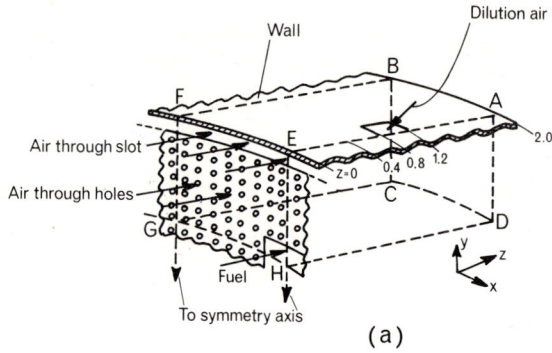

(a)

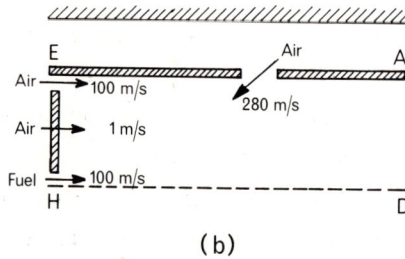

(b)

Fig. 13   The geometry of the gas-turbine combustor.

Figure 13 shows the geometry and some of the boundary conditions of the problem considered. The calculation domain ABCDEFGH can be considered to be a slice of an annular gas-turbine combustor. For convenience, the region is taken as exactly rectangular. The gaseous fuel enters through the region near H. Air enters through many small holes in the end plate EFGH and also through the slot near the line

EF. There is a further supply of air through the
"dilution-air hole" in the wall ABFE. The products
of combustion exit through the plane ABCD.

In Fig. 13(b), the values of the inlet velocities
are shown. All inlet temperatures are taken as
600°K. The wall is supposed to be cooled from outside by a stream of air at 600°K. A uniform heat-
transfer coefficient is assumed to govern the heat
transfer from the wall to the cooling air outside.
Unless otherwise stated, the results presented correspond to the heat-transfer coefficient of
300 J/s m^2 ^{O}K. The turbulence is calculated by the
two-equation model described in Sec. 4.1; the combustion is supposed to be diffusion-controlled; and
the six-flux radiation model outlined in Sec. 4.3 is
employed, with zero scattering and an absorptivity of
2 m^{-1}, except when the effect of various values of
absorptivity is investigated.

It may be useful here to enumerate the equations
and their dependent variables. We solve four hydrodynamic equations, namely the continuity equation and
the three momentum equations, to obtain the pressure
and the three velocity components. For turbulence,
two differential equations, namely those for k and
ε, are solved. The diffusion-flame approximation
makes it possible to obtain the concentrations of the
three species (fuel, oxygen, product) from the solution of the differential equation for the composite
variable m_{fu} - m_{ox} / i. Lastly, we add to the list
the equations for enthalpy and for the radiation quantities (I + J), (K + L) and (M + N). Thus, a
total of *eleven* differential equations are solved for
this problem. The computer time for a 7 × 7 × 7

S.V. Patankar

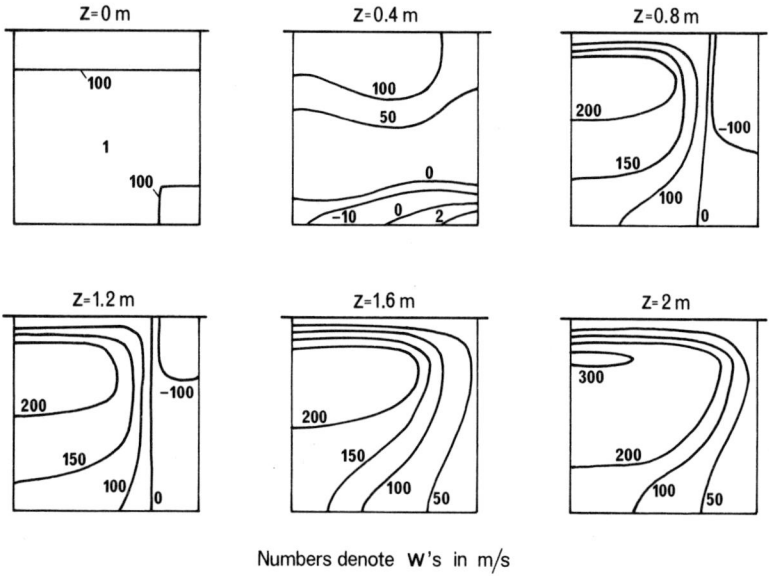

Z=0 m   Z=0.4 m   Z=0.8 m

Z=1.2 m   Z=1.6 m   Z=2 m

Numbers denote **w**'s in m/s

Fig. 14   Contours of the velocity w.

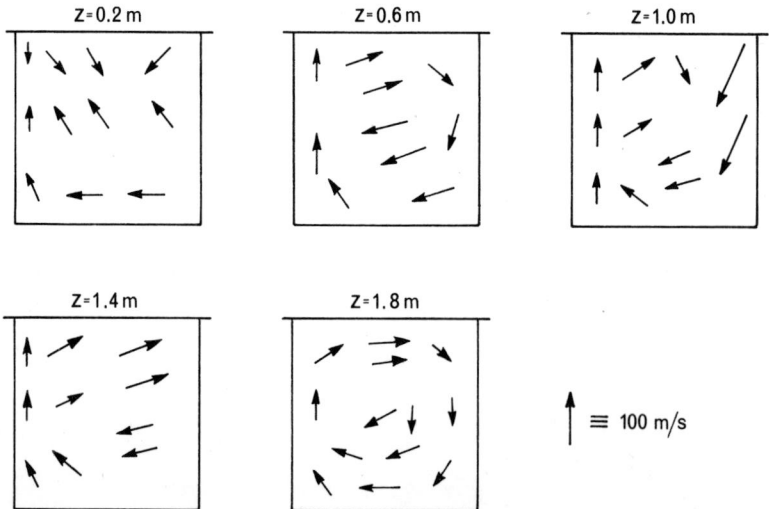

Z=0.2 m   Z=0.6 m   Z=1.0 m

Z=1.4 m   Z=1.8 m

$\uparrow \equiv$ 100 m/s

Fig. 15   The velocity field in xy planes.

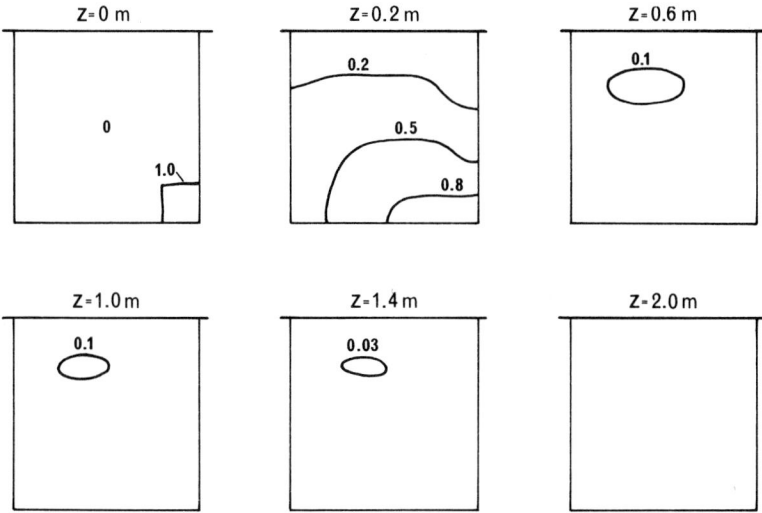

Fig. 16   Contours of the mass fraction of fuel.

grid is about 60 seconds on a CDC 6600.

Figures 14 to 19 show some of the results.   The
velocity distributions are presented in Fig. 14 and
15.   The velocity in the   z   direction is shown by
means of contours on various   xy   planes.   The magnitude of the velocity is seen to increase as we go
to the downstream end of the combustor;   this is
because the density has decreased as a result of combustion.   In most planes, there is a high-velocity
region near the top edge, which is due to the "film-
cooling" air injected along the wall.   Regions of
negative velocity can be seen;   they are caused
mainly by the fact that the dilution air is injected
with a strong upstream velocity component.   The velocity field in the   xy   planes is shown by velocity
vectors in Fig. 15.   A significant swirling motion

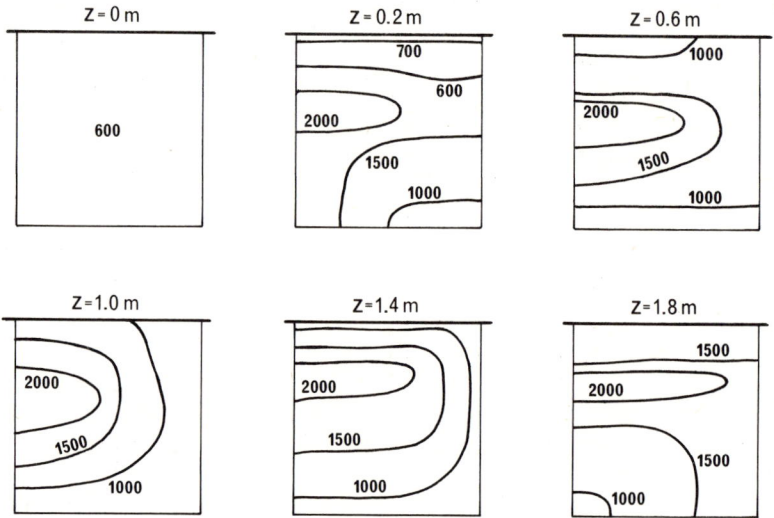

Fig. 17   Contours of the temperature of the fluid.

Section at z = 1 m

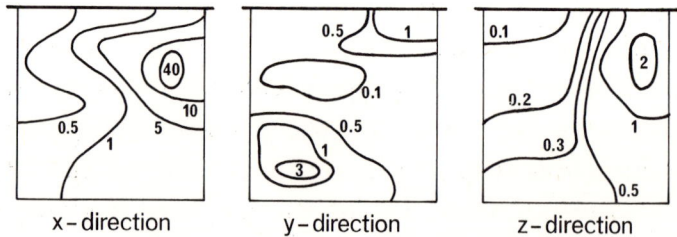

Fig. 18   Contours of the radiation flux-sums nor-
               malized by the local value of   2E.

is noticeable; this is also caused by the air injection through the dilution hole.

Figure 16, where the mass fraction of fuel is shown by contours, gives some idea of the flame shape. After its injection the fuel can be seen to spread over almost the whole cross-section of the combustor, but as mixing proceeds the unburnt fuel gets confined to a small region and then disappears completely.

Figure 17 exhibits the corresponding temperature distribution. The high-temperature regions can be seen where the chemical reaction takes place. The wall temperature is kept low by the air film blown along the wall. The temperature distribution at the outlet plane is very non-uniform; calculations of this kind could provide guidance to the designer on how to alter the geometry and other conditions to achieve a more uniform outlet-temperature distribution.

Figure 18 shows only a sample of the large amount of information generated by the radiation calculation. Contours are shown on one xy plane of the flux sums $(I + J)$, $(K + L)$ and $(M + N)$ for the three directions divided by the local value of 2E. If thermal equilibrium prevailed, these quantities would all be unity. Since the x-direction radiation does not see any cold surface, the radiation fluxes are the largest in the x-direction. On the other hand, the presence of the cooled wall makes the radiation fluxes smallest in the y direction.

Figure 19 is included as an example of a parametric study. Figure 19(a) presents the effect of the gas absorptivity on the wall temperature, and Fig. 19(b) shows the effect of the external heat-transfer coefficient. It can be seen that these

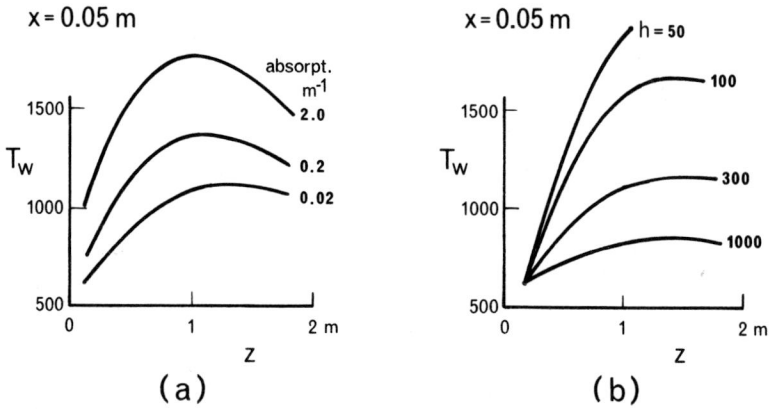

Fig. 19   Effect of the heat-transfer coefficient
               and the absorptivity on the wall tempera-
               ture.

results are in accordance with our qualitative expectations.   Perhaps these results may suffice to give
the reader a general idea of the complex computation.
More details and further predictions can be found in
the paper by Patankar and Spalding [22].

PART II.   THREE-DIMENSIONAL PARABOLIC FLOWS

6.   PARABOLIC FLOWS AND THEIR CALCULATION

6.1   Examples of Parabolic Flows

There are steady three-dimensional flows for
which we can find one co-ordinate in which the physical influences are experienced in only one direction.
As a result, the governing differential equations are
parabolic in one space co-ordinate.   Such flows are
sometimes called three-dimensional *parabolic* flows.

Flows of this kind derive their special character from the fact that there exists a predominant
direction of flow (there is no reverse flow in that
direction);   consequently, the upstream conditions
alone determine what happens downstream.   Thus parabolic flows include what are conventionally known as
boundary-layer flows, but also some other types as the
following examples will show.

Figure 20 shows four cases where the flow is
three-dimensional and can be regarded as parabolic.
In Fig. 20(a), we have a conventional boundary layer
on an aerofoil.   The three-dimensionality is caused
by the finiteness of the width of the aerofoil and by
the fact that the flow direction is not normal to the
leading edge.   If no separation occurs, the flow can
be treated as parabolic.

Figure 20(b) illustrates a free jet in stagnant
surroundings.   Here the three-dimensionality is due
to the square shape of the nozzle.   But, like the
round jet, the flow is determined by the upstream conditions at the nozzle exit.

(a)   (b)

(c)   (d)

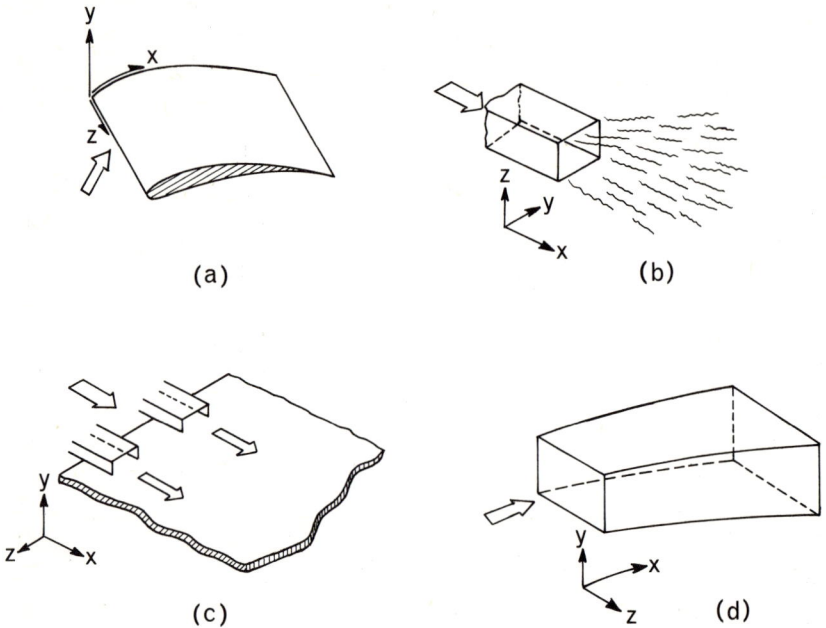

Fig. 20   Examples of three-dimensional parabolic
               flows.

In Fig. 20(c) we come to a wall-jet situation.
Once again the flow has one predominant direction.
The discreteness of the jets creates the three-
dimensionality.

All these situations involve unconfined flows.
In Fig. 20(d) we have an example of a confined flow.
The developing flow through a duct of non-circular
cross-section is three-dimensional, and provided that
no separation occurs it satisfies the requirements
for a parabolic flow.

6.2   *Available Calculation Procedures*

It goes without saying that the general calculation procedures referred to in Part I of this

article can be applied to three-dimensional parabolic
flows.   But that approach would ignore the special
character of these flows and would thus use too much
computer storage and time.   By taking advantage of
the parabolic character, a *marching* procedure may be
employed going from upstream to downstream.   That is,
at any given stage, we need to focus our attention on
only one cross-stream plane (i.e. a plane normal to
the predominant flow direction) and, having found the
values of variables on that plane, proceed to the
next plane downstream.   Thus, we need only a two-
dimensional storage and achieve a certain saving in
the computer time.

Most available procedures for three-dimensional
boundary-layer flows have been developed for a restricted class of parabolic flows.   They all seek to
predict the flow between an inviscid free stream and
a solid wall.   These procedures include the calculation methods by Raetz [23], Hall [24], Dwyer [25],
Fannelop [26], Nash [27], Krause and Hirschel [28],
Der [29], and East and Pierce [30].   These methods
assume that the stresses and diffusion fluxes are
significant only on the planes parallel to the wall,
and that the pressure does not vary in the direction
normal to the wall.   These assumptions make the
methods unsuitable for general parabolic flows.   They
are obviously inapplicable to confined flows like the
one in Fig. 20(d);   but they cannot give reasonable
results even for the situation in Fig. 20(c), where
it is the pressure variation in the cross-stream
plane that is responsible for producing the cross-
stream velocities.

In recent years, a few procedures have been developed which take into account the stresses and diffusion fluxes on planes normal to both the cross-stream directions and which take full account of the pressure variations in the cross-stream plane. These procedures are capable of predicting the general parabolic flows. Such procedures include the two methods described by Caretto, Curr and Spalding [31], the method by Patankar and Spalding [10] and a similar one by Briley [32]. One of the methods of Ref. [31] uses the velocity components and the pressure as the variables and treats them as simultaneous unknowns in a point-by-point substitution procedure; the other method suppresses the pressure and uses the vorticity in the main stream direction as a dependent variable. The method of Ref. [10] will be described in some detail in this article; indeed, historically, the procedure for elliptic flows described in Part I was derived as a generalization of the method of Ref. [10]. Reference [32] contains a procedure which is similar to that of [10] in many respects.

7.   THE CALCULATION PROCEDURE

7.1   *Governing Equations*

The equations that govern three-dimensional parabolic flows are simplified forms of the general equations presented in Part I. It is therefore sufficient to point out wherein this simplification lies. Let us assume that the predominant flow is in the x direction. Then, to turn the general differential equations into parabolic ones, we omit the viscous

stresses and diffusion fluxes acting on the yz plane.
In reality, these stresses and fluxes are very small
for the class of flows we are considering, and to re-
tain these quantities in our equations would allow
the downstream events to influence the upstream ones.

Thus the parabolic form of Eq. (3.1) for a
steady-state situation will be:

$$\text{div } (\vec{G}\phi) = \frac{\partial}{\partial y} \left(\Gamma_\phi \frac{\partial \phi}{\partial y}\right) + \frac{\partial}{\partial z} \left(\Gamma_\phi \frac{\partial \phi}{\partial z}\right) + S_\phi, \quad (7.1)$$

where the expression for the source term S_ϕ should
also be based on the parabolic-flow assumptions.

The special treatment given to the x coordinate necessitates another modification. We need to
"decouple" the pressure gradient that drives the main
stream velocity u from the pressure gradients for
the cross-stream velocities. It is therefore assumed
that the velocity u is influenced by a pressure $\bar{p}$
which is a form of space-averaged pressure over a
cross-stream plane. The variation of pressure *over*
that plane is given by a pressure p, which governs
the cross-stream velocities. Some justification of
this practice is given in Ref. [10]; here it may be
sufficient to record that the decoupling of pressures
mentioned above is essential to the success of a
marching solution procedure.

7.2 *Some Details of the Finite-difference Scheme*

Figure 21 shows portions of the two cross-stream
planes on which we concentrate during a forward step
in the marching process. From the knowledge of the
values of all the variables on the upstream plane,
our aim is to calculate the values on the downstream

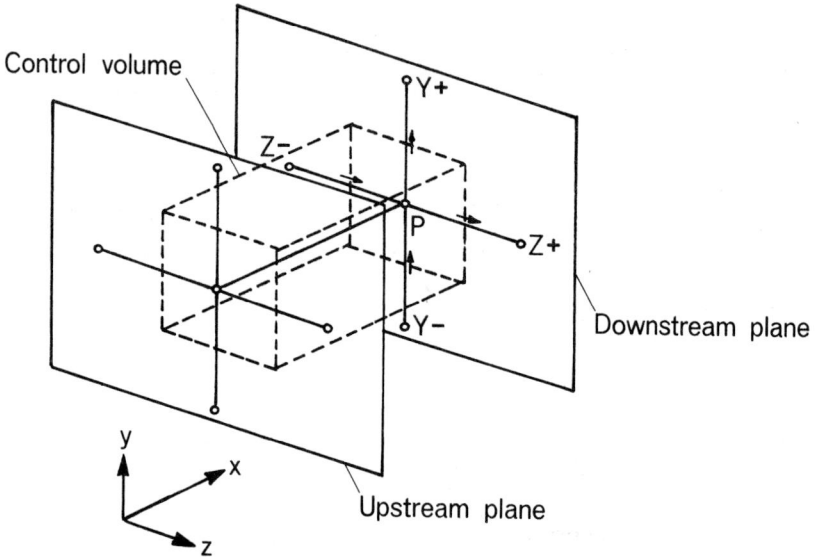

Fig. 21   The grid and the control volume for the
            parabolic calculation.

plane.   Thus, essentially we need to solve a two-
dimensional problem at each forward step.   It is
easy to imagine a two-dimensional version of the procedure described in Part I.   We form the difference
equations in much the same way.   The control volume
used for a general variable ϕ is shown in Fig. 21.
A fully-implicit scheme is used, which means the
downstream values of ϕ are assumed to prevail over
the control volume.

     The storage locations for the cross-stream vel-
ocity components v and w are staggered in the
same manner as in Part I; but, because of the
special treatment given to u, it is stored for the
node points for which all other variables are stored.

     The finite-difference equations resulting from
the momentum equations and from other conservation

equations have the form:

$$a_P \phi_P = a_{Y+} \phi_{Y+} + a_{Y-} \phi_{Y-} + a_{Z+} \phi_{Z+} + a_{Z-} \phi_{Z-} + b \, , \quad (7.2)$$

where b contains a contribution from the upstream (i.e. known) value of ϕ. Such equations are solved by the use of the TDMA traverses in the y and z directions.

The pressure $\bar{p}$ that governs the velocity u is determined in one of two ways: for unconfined flows, $\bar{p}$ is taken to be the same as the pressure prevailing in the free stream adjacent to the calculation domain; for confined flows, $\bar{p}$ is assumed to be uniform over a cross-section and is adjusted so that the resulting u's will satisfy the overall mass conservation for the duct.

Having obtained $\bar{p}$ in this manner and hence calculated the values of u for the downstream plane, we turn to the cross-stream velocities. First, a best estimate $p*$ of the pressure p is used to get the velocities $v*$ and $w*$ from the momentum equations. Then a pressure correction p' is proposed in the same manner as in Sec. 3.5, and, by the use of the local continuity equation, a pressure-correction equation is obtained, which has the same form as Eq. (7.2). The reason for getting this two-dimensional form rather than the three-dimensional form of Eq. (3.15) is that we do not allow the pressure p (and thus the correction p') to influence the velocity u.

The purpose of the above description has been to convey to the reader the basic ideas in specializing the procedure of Part I to three-dimensional para-

bolic flows.   Many algebraic details have been
omitted;   for these the reader may wish to turn to
Ref. [10].

7.3   Use of Other Co-ordinate Systems

So far, we have presented the calculation procedure by assuming that a Cartesian co-ordinate system is used.   Often there are advantages in the use
of other co-ordinate systems;   sometimes other systems are essential.   Some examples of these situations now follow.

The flow in a round pipe is normally regarded as
axisymmetrical (i.e. two-dimensional);   but, if the
inlet and boundary conditions do not possess axial
symmetry, a three-dimensional parabolic calculation
will be needed.   Then it would be more convenient to
use a cylindrical polar co-ordinate system.

If we consider the flow in a curved duct, it
will be appropriate to let the main-stream co-ordinate
x follow the curvature of the duct;   then account
must be taken of the fact that the distance Δx between two successive cross-stream planes varies over
the cross-section of the duct.

In a rectangular-sectioned diffuser, the cross-
stream calculation domains are rectangles, but their
dimensions vary with x.   In this case it is convenient to employ a grid in the $x \sim \eta \sim \zeta$ co-
ordinate system where η and ζ are y/y_D and
z/z_D respectively, y_D and z_D being the sides of
the cross-section of the diffuser.   This ensures
that η and ζ vary from 0 to 1 for any cross-
section considered.

For the case of the diffuser, y_D and z_D were
determined by its geometry. For external boundary
layers, we can choose y_D and z_D (and adjust them
appropriately at each forward step) such that the
flow region of interest is just contained within the
0-to-1 range of η and ζ. This is the device
used in the two-dimensional boundary-layer procedure
of Patankar and Spalding [33] and has applicability
for three-dimensions also.

8. SOME PREDICTIONS OF THREE-DIMENSIONAL PARABOLIC FLOWS

The use of the calculation procedure just described will now be illustrated in this section.
Four sets of calculations are presented; the first
three are for confined flows, while the last one re-
lates to an external flow. The situations chosen
happen to be such that none of them (not even the
last one) can be predicted by the use of "conventional" boundary-layer methods. All the results
presented here are for turbulent flow, and, in most
places, they are compared with experimental data.
Only the work reported in Sec. 8.4 below has so far
been published. Fuller accounts of the other flow
predictions are, however, expected to become available during 1974.

8.1 *Developing Turbulent Flow in a Square Duct*

The developing flow in a duct of non-circular
cross-section is probably the simplest example of a
three -dimensional parabolic flow, and, when the

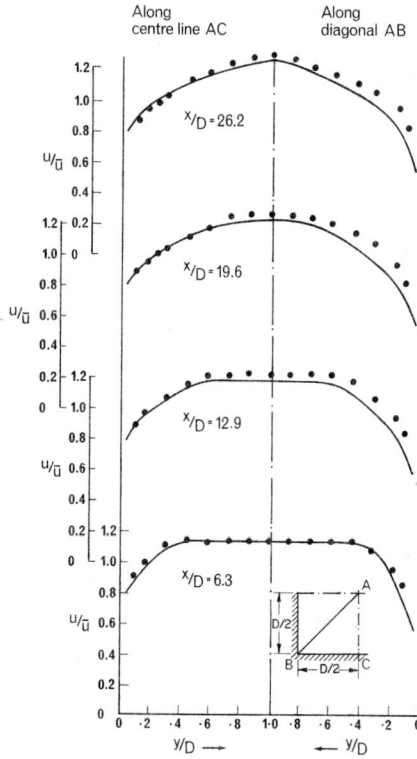

Fig. 22   Profiles of the longitudinal velocity in
           the turbulent flow in a square duct.

cross-section is a square, the computation becomes
especially straightforward.   The work presented here
is performed by Tatchell [34], who predicted the
developing region of the turbulent flow in a square
duct.

    The use of the two-equation turbulence model de-
scribed in Sec. 4.1 (or of any other model which uses
the turbulent-viscosity concept) is not entirely

satisfactory for the square-duct flow.   For, the
model will predict that, when the longitudinal velocity u becomes independent of the distance x
(the "fully-developed" condition), the cross-stream
velocities v and w fall to zero; in reality, the
experiments show that a secondary flow exists even in
the fully-developed state.   This secondary flow is
supposed to be driven by turbulent normal stresses
which do not strictly obey Eq. (4.1).   Launder [35]
proposed some algebraic formulae to calculate the
Reynolds stresses, which can remove this deficiency
of a two-equation model.   It is this modification to
the two-equation model of Sec. 4.1 that is used for
the computations presented in this sub-section.

Figure 22 shows by full lines the predicted variation of the longitudinal velocity u at different
distances downstream of the inlet section, at which
the velocity is uniform.   The dots show the experimental data of Ahmed and Brundrett [36].   It can be
concluded that the agreement of the predictions with
the experimental data is satisfactory.

8.2   *Developing Turbulent Flow in a Rectangular Diffuser*

Now we present some computations by Sharma [37]
for the case of a developing flow in a diffuser of
rectangular cross-section.   Unlike the case in Sec.
8.1 above, the inlet velocity distribution is non-
uniform.   The $x \sim \eta \sim \zeta$ co-ordinate system indicated in Sec. 7.3 is used so that the grid exactly
fits the diffuser cross-section at all longitudinal
locations.   The effect of turbulence is calculated
by the generalized mixing-length formula given by
Eq. (4.2) and the variation of the mixing length L

S.V. Patankar

(a)

(b)

(c)

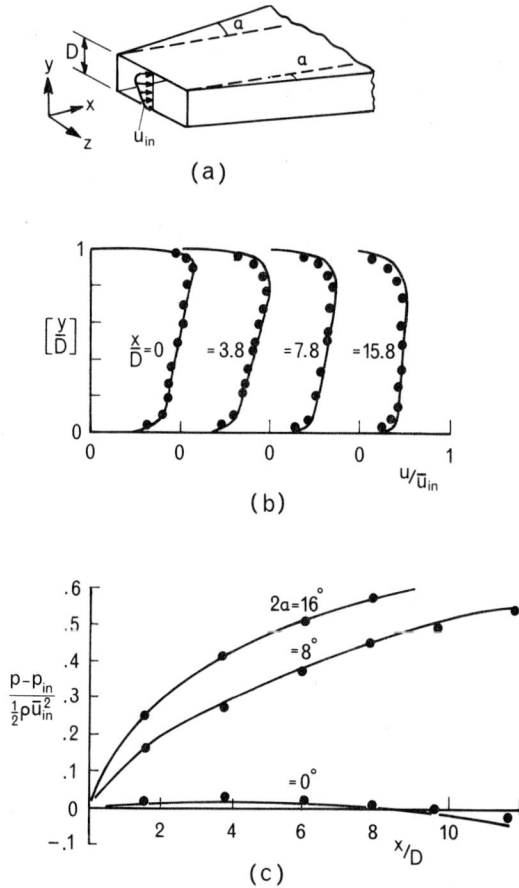

Fig. 23   Flow in a rectangular diffuser: velocity
            profiles and pressure variation.

is calculated according to the procedure given by
Buleev [17].

     Figure 23(a) shows the geometry and the co-
ordinate system used.   The predictions for the development of the longitudinal velocity u   are shown
in Fig. 23(b) by full lines;   the dots represent the
experimental data of Masuda *et al.* [38].   In Fig.

23(c), the variation of pressure with longitudinal
distance is presented for three different divergence
angles of the diffuser;   again the full lines and the
dots represent the predictions and the experimental
data respectively.

The agreement of theory and experiment can be
regarded as quite good from Figs. 23(b) and (c).

8.3   *A Transverse Turbulent Jet in a Duct*

Now we look at the situation in which a flow between two parallel plates is modified by a row of
round jets injected normally from one of the plates.
The geometry is shown in Fig. 24(a);   the results
plotted in Fig. 24(b) are based on the work of
Tatchell [39].

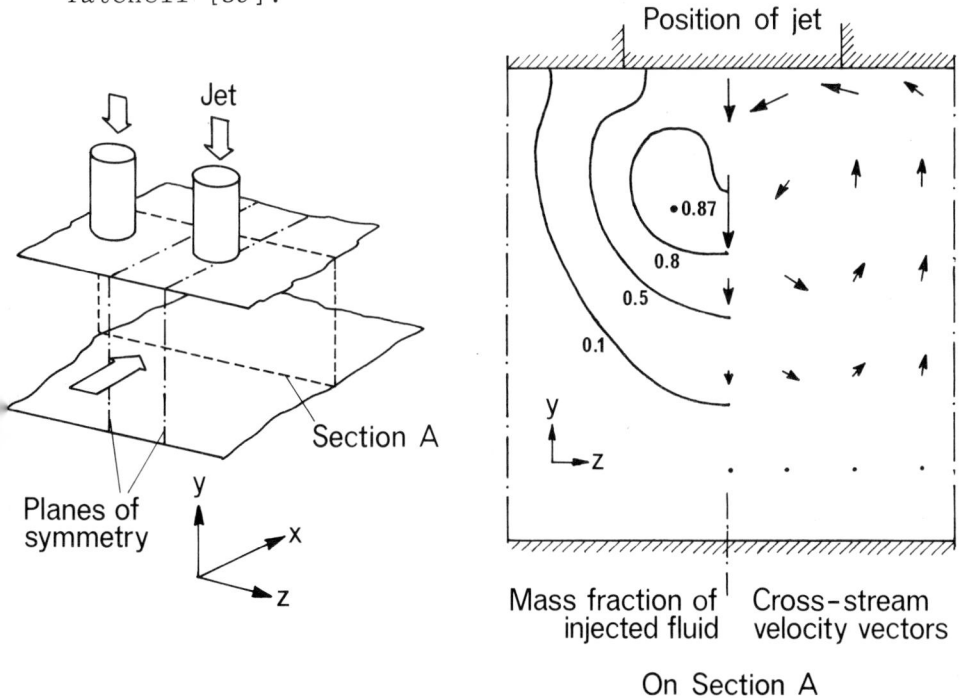

Fig. 24   A duct flow with a transverse jet.

If the injection velocity is large compared with
the longitudinal velocity, a significant recirculation
region will form immediately downstream of the jet.
Then our parabolic-flow procedure will not be applicable.   The results in Fig. 24(b) are obtained for
the case in which the jet velocity is equal to the
mean longitudinal velocity;   the calculation shows no
recirculation in the main stream direction.

The physical basis of the calculation is the two-
equation turbulence model outlined in Sec. 4.1.   No
comparisons with experimental data are made.

Fig. 24(b) shows the distributions over a cross-
stream plane just downstream of the jet.   One half
of the figure is used to show the cross-stream velocity field in the form of arrows.   The vortex set
up by the injected jet can be clearly seen.   In the
other half of the figure, contours of the mass fraction of the injected fluid are shown.   The kidney
shape of these contours is in accordance with how
jets deflected by a normal stream are known to behave.
It is also interesting to note that the maximum concentration does not lie on the centre line.

8.4 *Film Cooling by Means of a Row of Coolant Jets*

Lastly we present some results from the paper by
Patankar, Rastogi and Whitelaw [40], where more results and complete details can be found.   The geometrical situation is shown in Fig. 25.   A uniform
main stream flows over a smooth flat surface which is
cooled by a number of discrete round jets injected
tangential to the surface.   The constructional details of the coolant holes would normally create a
small recirculation zone in the vicinity of the holes;

Main stream

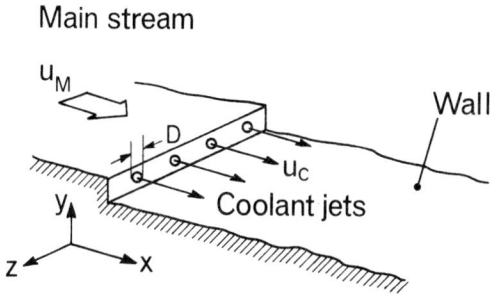

Fig. 25   The film-cooling situation considered.

this recirculation is ignored in the present computation.   A large number of flows with different jet-
to-mainstream velocity ratios and density ratios are
considered.   Computations are also performed for
those cases where the coolant is injected from only
one hole.   The predictions are compared with experimental data.

For an efficient use of the finite-difference
scheme, the $x \sim \eta \sim \zeta$ co-ordinate system is used.
In the case of a row of coolant jets, the z-direction
width of the calculation domain is kept uniform, but
the grid is allowed to stretch in the   y   direction
as the region of significant velocity variation
grows.   When only a single coolant jet is considered,
the grid expands in both   y   and   z   directions.   The
rate of grid expansion is decided by means of a so-
called entrainment formula in a manner similar to the
practice in Ref. [33].

The turbulence calculation is based on the generalized mixing-length formula (Eq. (4.2)).   The
variation of the mixing length   L   is assumed to be
that which is common for two-dimensional boundary
layers;   that is,   L   is taken as proportional to the

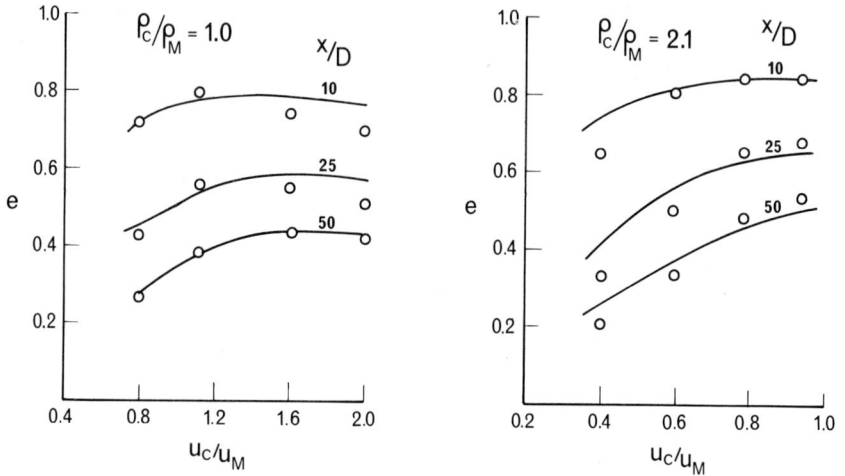

Fig. 26   Effectiveness for the multi-hole geometry.

distance from the wall until it attains the value
appropriate to the outer layer;   thereafter   L   is
taken to be uniform in the outer region.

In presenting the results, we shall concentrate
attention on the "effectiveness" of the film cooling,
e,   along the line directly in front of the centre of
a coolant jet.   The effectiveness is a dimensionless
measure of the temperature difference between the wall
and the main stream.   Some results for the multiple-
jet situation are shown in Fig. 26, and the results
for the single-jet case are given in Fig. 27.
Results in Fig. 27 are for uniform-density situation;
in Fig. 26, density ratios of   1   and   2.1   are used.
Both figures contain predictions and data for a number
of velocity ratios between the coolant jet and the
main stream.   Thus, in these two figures, a wide range
of operating conditions is covered.   The agreement
of the predictions with experimental data is quite good.

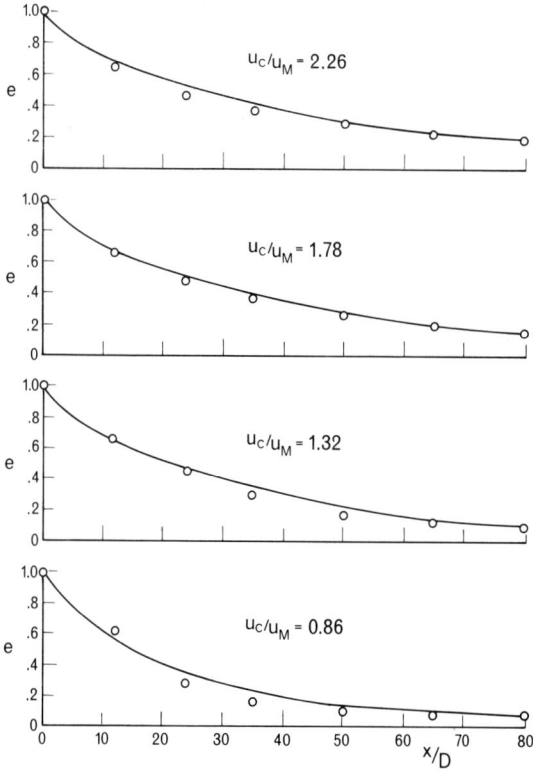

Fig. 27   Effectiveness for the single-hole geometry.

9.   CONCLUDING REMARKS

The present article has described a general calculation procedure for three-dimensional elliptic
flows and its simplified version for three-dimensional
parabolic flows.   A general framework was erected for
accommodating all the physical processes that are important in practical problems.   We have also shown

how some of these processes can be modelled. Examples
of calculations presented for both kinds of flows perhaps indicate to the reader that it is now feasible to
make predictions of practically-important phenomena.

This is an appropriate place to summarize the
important features of the method.   We use a staggered
grid for the velocity components which enables us to
use the pressure difference between adjacent nodes as
the driving force for the velocity component between
them.   As far as possible, we use an implicit scheme
and achieve the consequent economy.   Further economy
accrues from our use of the line-by-line, rather than
point-by-point, procedure for solving the algebraic
equations.   The use of the "hybrid" formula to represent the combined convection-diffusion effect gives
accuracy and ensures the convergence of the iterative
process of solving the linear difference equations.
The possibility of divergence resulting from the non-
linearities is prevented by appropriate underrelaxation.   In the parabolic-flow procedure, the decoupling of pressures cuts off the elliptic influences (i.e. propagation of the downstream effects to
upstream locations) that would otherwise make the
marching calculation impossible.

Although significant progress has already been
made towards the development of satisfactory means of
predicting practical phenomena, further work of two
kinds is still needed.   The first concerns the organization of a computer program such that it is able to
accommodate, with the help of the secondary storage
in a computer, a grid of arbitrary fineness.   When
that is done, it will be possible to represent in a
calculation all the details of a practical geometry

and to ensure that the results are independent of the grid fineness.   The second kind of further work needed is regarding the mathematical models for various physical processes of interest.   Existing models should be tested and improved before they can be accepted as reliable for design purposes.

In the past few years, procedures and models for two-dimensional parabolic flows have been brought to a fairly high level of reliability;   there are signs of success in the prediction of complex two-dimensional elliptic flows;   it therefore may not be long before we can make reliable predictions of three-dimensional flow phenomena.

ACKNOWLEDGEMENTS

The work described in this paper has been performed at Imperial College, London by a group of research workers, for whom Professor D.B. Spalding has provided guidance and inspiration.   Thanks are also due to Mr. D. Sharma and Mr. D.G. Tatchell for making available to me some of their unpublished results.

REFERENCES

1.   Aziz, K. and Hellums, J.D.   "Numerical solution of the three-dimensional equations of motion for laminar natural convection". *Physics of Fluids*, <u>10</u>, p. 314, 1967.

2.   Mallinson, G.D. and de Vahl Davis, G.   "The method of the false transient for the solution of coupled elliptic equations".   Univ. of New South Wales, Kensington, Australia, Report No. 1972/FMT/2, 1972.

3.   Harlow, F.H. and Welch, J.E.   "Numerical calcu-
     lation of time-dependent viscous incompress-
     ible flow of fluid with free surface".
     Physics of Fluids, <u>8</u>, p. 2182, 1965.

4.   Thommen, H.U.   "Numerical integration of the
     Navier-Stokes equations".   *ZAMP*, <u>17</u>, p. 369,
     1966.

5.   Chorin, A.J.   "Numerical solution of the Navier-
     Stokes equations".   *Math. Computation*, <u>22</u>,
     p. 745, 1968.

6.   Williams, G.P.   "Numerical integration of the
     three-dimensional Navier-Stokes equations for
     incompressible flow".   *J. Fluid Mech.*, <u>37</u>,
     p. 727, 1969.

7.   Deardorff, J.W.   "A numerical study of three-
     dimensional turbulent channel flow at large
     Reynolds numbers".   *J. Fluid Mech.*, <u>41</u>,
     p. 453, 1970.

8.   Amsden, A.A. and Harlow, F.H.   "The SMAC method:
     a numerical technique for calculating incom-
     pressible fluid flows".   Los Alamos Scien-
     tific Laboratory, LA-4370, 1970.

9.   Zuber, I.   "Ein mathematisches Modell des
     Brennraums".   Monographs and Memoranda no.
     12.   Staatliche Forschungs Institut für
     Maschinenbau.   Bechovice, Czechoslovakia,
     1972.

10. Patankar, S.V. and Spalding, D.B.   "A calcu-
     lation procedure for heat, mass and momentum
     transfer in three-dimensional parabolic
     flows".   *Int. J. Heat Mass Transfer*, <u>15</u>,
     p. 1787, 1972.

11. Caretto, L.S., Gosman, A.D., Patankar, S.V. and
     Spalding, D.B.   "Two calculation procedures
     for steady, three-dimensional flows with re-
     circulation".   *Proceedings of Third Int.
     Conf. on Num. Methods in Fluid Dynamics*,
     Paris, July 1972.

12. Patankar, S.V. and Spalding, D.B.   "Numerical
     predictions of three-dimensional flows".
     Imperial College, Mech. Eng. Dept., Report
     EF/TN/A/46, June 1972.

13. Spalding, D.B.   "A novel finite-difference for-
     mulation for differential expressions involv-

ing both first and second derivatives". *Int. J. Num. Methods in Eng.*, <u>4</u>, p. 551, 1972.

14. Patankar, S.V. "Calculation of unsteady compressible flows involving shocks". Imperial College, Mech. Eng. Dept., Report UF/TN/A/4, December 1971.

15. Launder, B.E. and Spalding, D.B. "Mathematical models of turbulence". Academic Press, London, 1972.

16. Launder, B.E. and Spalding, D.B. "Turbulence models and their application to the prediction of internal flows". University of Salford, Symposium on Internal Flows, paper 1, 1971.

17. Buleev, N.I. "Theoretical model of the mechanism of turbulent exchange in fluid flows". Teploperedacha, USSR Academy of Sciences, Moscow, p. 64, 1962.

18. Spalding, D.B. "Mixing and chemical reaction in steady confined turbulent flames". Thirteenth Symposium on Combustion, Salt Lake City, 1970.

19. Mason, H.B. and Spalding, D.B. "Prediction of reaction rates in turbulent pre-mixed boundary-layer flows". Imperial College, Mech. Eng. Dept. Report HTS/73/11, February 1973.

20. Patankar, S.V. and Spalding, D.B. "A computer model for three-dimensional flow in furnaces". Fourteenth Symposium on Combustion, August 1972.

21. Patankar, S.V. and Spalding, D.B. "A calculation procedure for the transient and steady-state behaviour of shell-and-tube heat exchangers". *Proceedings of the Int. Summer School in Yugoslavia*, September 1972.

22. Patankar, S.V. and Spalding, D.B. "Simultaneous predictions of flow pattern and radiation for three-dimensional flames". International Seminar on "Heat Transfer from Flames" in Yugoslavia, 1973.

23. Raetz, G.S. "A method of calculating three-dimensional laminar boundary layers of steady compressible flows". Northrop Aircraft Inc., Rep. No. NAI-58-73 (BLC-144), 1957.

24. Hall, M.G.   "A numerical method for calculating steady three-dimensional laminar boundary layers".   RAE Tech. Rep. 67145, June 1967.

25. Dwyer, H.A.   "Solution of a three-dimensional boundary layer flow with separation".   *AIAA J.*, <u>6</u>, p. 1336, 1968.

26. Fannelop, T.K.   "A method of solving the three-dimensional laminar boundary-layer equations with application to a lifting re-entry body".   *AIAA J.*, <u>6</u>, p. 1075, 1968.

27. Nash, J.F.   "The calculation of three-dimensional turbulent boundary layers in compressible flow".   *J. Fluid Mech.*, <u>37</u>, p. 625, 1969.

28. Krause, E. and Hirschel, E.H.   "Exact numerical solutions for three-dimensional boundary layers".   Second Int. Conf. on Num. Methods in Fluid Dynamics, Univ. of California, Berkeley, September 1970.

29. Der, J.   "A study of general three-dimensional boundary-layer problems by an exact numerical method".   *AIAA J.*, <u>9</u>, p. 1294, 1971.

30. East, J.L. and Pierce, F.J.   "Explicit numerical solution of the three-dimensional incompressible turbulent boundary-layer equations".   *AIAA J.*, <u>10</u>, p. 1216, 1972.

31. Caretto, L.S., Curr, R.M. and Spalding, D.B.   "Two numerical methods for three-dimensional boundary layers".   *Comp. Methods in Appl. Mech. and Eng.*, <u>1</u>, p. 39, 1972.

32. Briley, W.R.   "The computation of three-dimensional viscous internal flows".   *Proceedings of Third Int. Conf. on Num. Methods in Fluid Dynamics*, Paris, Vol. II, July 1972.

33. Patankar, S.V. and Spalding, D.B.   "Heat and mass transfer in boundary layers".   Intertext Books, London, Second Edition, 1970.

34. Tatchell, D.G.   "Developing turbulent flow in a square duct".   Unpublished work at Imperial College, London, 1972.

35. Launder, B.E.   "An improved algebraic modelling of the Reynolds stresses".   Imperial College, Mech. Eng. Dept. Report TM/TN/A/9, April 1971.

36. Ahmed, S. and Brundrett, E. "Turbulent flow in non-circular ducts. Part 1: mean flow properties in the developing region of a square duct". *Int. J. Heat Mass Transfer*, **14**, p. 365, 1971.

37. Sharma, D. "Developing turbulent flow in a rectangular diffuser". Unpublished work at Imperial College, London, 1972.

38. Masuda, S., Arida, I., and Watanabe, I. "On the behaviour of uniform shear flow in diffusers and its effects on diffuser performance". *J. Eng. Power*, **93A**, p. 377, 1971.

39. Tatchell, D.G. "Flow in a duct with a transverse jet". Unpublished work at Imperial College, London, 1972.

40. Patankar, S.V., Rastogi, A.K. and Whitelaw, J.H. "The effectiveness of three-dimensional film-cooling slots - predictions". *Int. J. Heat Mass Transfer*, **16**, p. 1673, 1973.

NOMENCLATURE

Symbol	*Meaning*
a	(without subscript) absorptivity for radiation
a	(with subscript) a coefficient in the finite-difference equation
A_x, A_y, A_z	areas of control-volume faces normal to the x, y and z directions respectively
b	the term (in the finite-difference equation) which does not contain the dependent variable
c_1, c_2, c_D	constants in the turbulence model
C_{x+}	convection and diffusion flux in the x direction
d_{x-} D_{x-}	multipliers of the pressure-difference term in the momentum equations
D	a typical dimension of the cross-section of a duct

Symbol	Meaning
e	effectiveness of film cooling ($\equiv (T_W - T_M)/(T_C - T_M)$)
E	the black-body emissive power at the gas temperature
g	the generation rate of turbulence energy
$\vec{G}$	the mass-flux vector ($\equiv \rho\vec{V}$)
h	the external heat-transfer coefficient
i	the ratio of the mass of oxygen to that of fuel in a stoichiometric mixture
$\vec{i}$	unit vector in direction i
I	radiation flux in the positive x direction
J	radiation flux in the negative x direction
k	kinetic energy of turbulence per unit mass
K	radiation flux in the positive y direction
ℓ	the length scale of turbulence
L	the mixing length, or the radiation flux in the negative y direction
m_{fu}	mass fraction of fuel
m_{ox}	mass fraction of oxygen
M	radiation flux in the positive z direction
N	radiation flux in the negative z direction
p	pressure
p^*, p', $\bar{p}$	estimated pressure, pressure correction and mean pressure respectively
s	scattering coefficient
S_ϕ	source term for the variable ϕ
S_i	source term (excluding the pressure gradient) for the i-direction momentum

Symbol	Meaning
t	time
t_{res}	residence time of the tube fluid in the heat exchanger
T	temperature
T_C, T_M, T_W	temperature of coolant, main stream and wall respectively
u	velocity in the x direction
u^*	u based on the estimated pressure
u'	turbulent fluctuation of u
$\bar{u}$	average value of u for a cross-section
u_C, u_M	velocity of coolant and main stream respectively
v	velocity in the y direction
v^*	v based on the estimated pressure
V_i	velocity component in the i direction
$\vec{V}$	the velocity vector
w	velocity in the z direction
w^*	w based on the estimated pressure
x	a co-ordinate; the predominant flow direction for parabolic flow
x_1, x_2, x_3	co-ordinates when tensor notation is used
y	a co-ordinate
y_D	the y-direction width of the calculation domain
z	a co-ordinate
z_D	the z-direction width of the calculation domain
Γ_t	turbulent exchange coefficient
Γ_v	exchange coefficient for momentum; i.e. viscosity

Symbol	*Meaning*
Γ_ϕ	exchange coefficient for the variable ϕ
δ_{ij}	Kronecker delta (= 1 for i=j; =0 for i≠j)
$\delta x+$	the x-direction distance between two nodes
Δt	the time step
Δx, Δy, Δz	lengths of sides of the control volume
ε	kinematic dissipation rate of turbulence energy
ζ	normalized co-ordinate ($\equiv z/z_D$)
η	normalized co-ordinate ($\equiv y/y_D$)
μ_t	turbulent viscosity
ρ	density
ρ_C, ρ_M	density of coolant and main stream respectively
σ_k	Prandtl number for k transport
σ_t	turbulent Prandtl number
σ_ε	Prandtl number for ε transport
ϕ	a general dependent variable

Subscript	*pertaining to*
P, x+, x-, X+, X-, y+, y-, Y+, Y-, z+, z-, Z+, Z-	the locations shown in Figs. 1, 2 and 3

Superscript	*pertaining to*
o	the (known) value of a variable at time t
———	time-averaged value

A REVIEW OF EXPERIMENTAL DATA OF UNIFORM DENSITY
FREE TURBULENT BOUNDARY LAYERS

by

W. Rodi

Sonderforschungsbereich 80, University of Karlsruhe
Germany

ABSTRACT

Experimental data of the following free turbulent
boundary layers are critically reviewed: mixing
layers, plane, round, and radial jets, the former two
issuing into both stagnant and moving surroundings,
plane and round wakes, and shear-free layers behind
self-propelled bodies. Attention is focussed on the
more recent data and on those experimental quantities
that are of relevance to present-day turbulence modelling. These quantities are identified by considering the exact equations for momentum, turbulent shear-
stress, and turbulent kinetic energy. The following
measurements are reviewed: rate of spread, longitudinal mean velocity, turbulent shear-stress and kinetic
energy, and energy and shear-stress balances. The
experimental results are compared with theoretical
expectations, particularly concerning the similarity
behaviour. To enable the comparison, the theoretical
similarity laws are presented for all potentially
self-similar free boundary layers. The measurements
of various experimenters are compared and examined for
their reliability; and, mainly on the basis of consistency checks, the most reliable data are identified
for each flow case.

1.   INTRODUCTION

Importance of experimental data.   The basic objective of any study of turbulence, whether theoretical or experimental, is to contribute to the development of a general calculation method for turbulent
flows.   The contribution of experiments may be described as follows.   With present computers, we cannot solve the full time-dependent Navier-Stokes
equations governing turbulent flows.   Time-averaged
equations are therefore used in engineering calculations; but these lack some of the information contained in the Navier-Stokes equations.   This information must be provided by experiments; and the
usual way to obtain a closed set of time-averaged
equations is to devise a turbulence model describing
the apparent (or Reynolds) stresses.*   Experiments
play two important roles in the modelling process.
Firstly, they are the basis of our understanding of
turbulence phenomena and thus guide the formulation
of a model.   Secondly, the empirical constants and
functions contained in the model can be determined
only from experiments.   For most models, these constants and functions are likely to be valid for a
restricted range of flows only (since the time-
dependent Navier-Stokes equations are the only truly
universal "model"); and to find out this range, experiments are needed once more.   Thus, we may con-
clude, experiments are vital to the development of
calculation methods for turbulent flows.

---

* For an introduction to turbulence modelling see
Launder and Spalding [1].

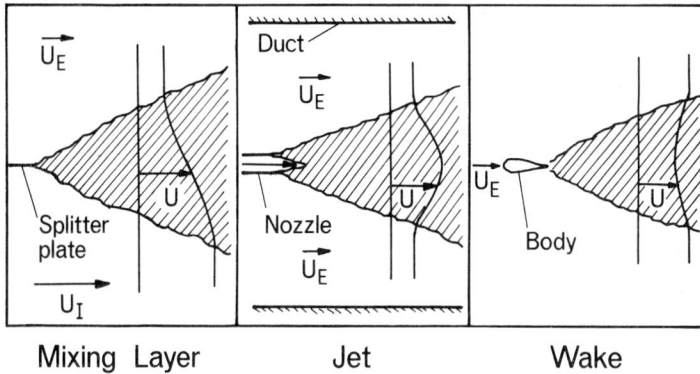

Mixing Layer   Jet   Wake

Fig. 1   The three basic examples of free turbulent
boundary layers.

The flows considered.   This article is concerned with experimental data for a class of flows
called "free turbulent boundary layers".   The label
"free" implies that the flows are remote from walls;
and the label "boundary layer" is attached to those
flow regions in which there is a single predominant
flow direction, and in which shear stresses and diffusion fluxes are significant only in directions per-
pendicular to the predominant direction.   Three
typical examples of free turbulent boundary layers
are sketched in Fig. 1:   the mixing layer, the jet,
and the wake.   All the flows considered in the article
are variations or combinations of these three basic
examples.*   Both plane and axisymmetric (without

---

*   The combination of jet and wake with zero momentum
excess (self-propelled-body flow) is included.   In the
far field of this flow, the shear stress is zero.   For
this reason, the name "free boundary layer" is preferred to the more common name "free shear layer".

swirl) situations are included.   The article is restricted to flows with steady mean quantities, high
Reynolds number, and constant density.   The shaded
areas in Fig. 1 define the free turbulent boundary
layer regions.   The flow outside these regions is
called free stream and is normally considered as uniform and non-turbulent.

Free turbulent boundary layers are an important
group of flows for two reasons.   Firstly, they play
a significant role in many kinds of engineering equipment, in rivers, and in the atmosphere: jet engines,
jet ejectors, wakes behind aeroplanes and submarines,
cooling water dispersal in rivers, chimney plumes,
and jet streams in the atmosphere are a few examples.
Secondly, free turbulent boundary layers, despite
their relative simplicity, exhibit some features
which make them crucial test cases for any general
prediction method.   For example, they include significant regions with both strong and weak strain and,
owing to this feature, are more sensitive to the modelling of turbulence* than are wall boundary layers
which are controlled largely by the strong-strain near-
wall region.   It will be recognized, therefore, that
experimental data for free turbulent boundary layers
are needed greatly, both for the development of calculation methods for these flows and for the development
of more general methods.

Purpose of the study.   Free turbulent boundary
layers have been measured fairly extensively;   so
there are plenty of data available for most laboratory

---

* For a comparison of the performance of six turbulence models in free shear layers see Launder *et al*. [2].

flow situations of interest.   However, the agreement
among experiments of nominally identical flows is
rather poor in some cases, particularly so for the
turbulence data;   and it is these data that become increasingly important as the turbulence models become
more complex.   It appears very important to find and
use the most reliable data;   for a model cannot be
better than its empirical input.   However, researchers
will find it very time-consuming to work through all
the original literature and to check the reliability
of all the data.   Existing reviews* are restricted
to earlier data and deal mostly with mean flow quantities only.   It is the purpose of this study to review the more recent data on free turbulent boundary
layers, including turbulence data, to check their reliability, and on the basis of this check, to identify
the most reliable data for each flow case.   In a
later publication, the data will be used to verify a
turbulence model developed by the author (see also
Rodi [16]).

  Outline of the paper.   A great many quantities
can be and have been measured in free turbulent boundary layers;   but only those will be included in the
review that are relevant to present-day turbulence
modelling.   By reference to current models, Sec. 2
identifies these quantities.   Section 3 presents
some theoretical considerations;   these facilitate
the testing of the data for internal consistency and
they lead to the expectation that many of the flows

---

* Readers interested in reviews of earlier data are
referred to Abramovich [3], Halleen [4], Spalding [5],
Newman [6], and Harsha [7].

to be considered are self-similar.   Also given in
Sec. 3 are the decay laws expected in these cases.
Section 4 states the criteria used for the selection
of the most reliable data.   The actual review starts
in Sec. 5.   There the following potentially self-
similar flows are dealt with: shear-free layers,
mixing layers, plane, round, and radial jets, and
plane and round wakes.   Non-similar jets (in a uniformly moving stream) are reviewed in Sec. 6.
Measurements of the following quantities are docu-
mented and discussed: rate of spread, profiles of
velocity, shear stress, and turbulent kinetic energy,
and energy and shear-stress balances.   The measurements are compared with the expectations from the
theory;   and possible explanations are offered for any
lack of agreement that is found.   Finally, in Sec. 7
some recommendations are made for future experiments.

2.   THE DATA REQUIRED

Engineers mainly require of a calculation method
for turbulent flow that it should predict mean-flow
quantities correctly.   Hence, these quantities are
the main "target" values we want to extract from experiments.   However, measurements of turbulence quantities can aid greatly the construction of turbulence
models;   so they are essential too.   In order to
establish the quantities relevant to turbulence modelling, we consider briefly the equations on which many
current approaches are based.

2.1   Some Basic Equations of Turbulence Modelling

The aim of any prediction method is to solve the momentum equation

$$U \frac{\partial U}{\partial x} + V \frac{\partial U}{\partial y} \ = \ U_E \frac{dU_E}{dx} - \frac{1}{y^j} \frac{\partial}{\partial y} (y^j \ \overline{uv}) - \frac{\partial}{\partial x} (\overline{u^2} - \overline{v^2}),$$

$$(2.1)$$

together with the continuity equation

$$\frac{\partial U}{\partial x} + \frac{1}{y^j} \frac{\partial}{\partial y} (y^j \ V) \ = \ 0 \ . \qquad (2.2)$$

The equations are written in a form valid for two-dimensional (including axisymmetric) boundary-layer flows at high Reynolds numbers.   The symbols are defined in the nomenclature.   As mentioned in the Introduction, a turbulence model is necessary to describe the Reynolds stresses* appearing in the momentum equation.   These stresses are governed by transport equations which can be derived in exact form from the Navier-Stokes equations.   In fact most physical models for calculating turbulent flows may be regarded as approximated forms (sometimes drastically approximated) of these stress transport equations.   In two-dimensional boundary layers, the shear stress $\overline{uv}$ is the only Reynolds stress that has a significant effect on the mean flow, so the $\overline{uv}$-equation is of particular importance.   Also, for these flows, the normal stresses are usually considered only in the form of half their sum, which is the turbulent

---

* In equation (2.1), these stresses are $\overline{uv}$, $\overline{u^2}$, and $\overline{v^2}$.

W. Rodi

kinetic energy, k (*e.g.* Hanjalić and Launder [8]). The equations for $\overline{uv}$ and k valid for boundary layers at high Reynolds numbers read (Rodi [9]):

$$U \frac{\partial \overline{uv}}{\partial x} + V \frac{\partial \overline{uv}}{\partial y} = \underbrace{- \frac{1}{y^j} \frac{\partial}{\partial y} [y^j \, \overline{u(v^2 + p/\rho)}] + j \frac{\overline{uw^2}}{y}}_{}$$

$$\underbrace{\phantom{U \frac{\partial \overline{uv}}{\partial x}}}_{\text{convection}} \qquad \underbrace{\phantom{- \frac{1}{y^j} \frac{\partial}{\partial y} [y^j \, \overline{u(v^2 + p/\rho)}] + j \frac{\overline{uw^2}}{y}}}_{\text{diffusion}}$$

$$\underbrace{- \overline{v^2} \frac{\partial U}{\partial y} + j\overline{uv} \frac{V}{y}}_{\text{production}}$$

$$\underbrace{+ \frac{p}{\rho} \left(\frac{\partial u}{\partial y} + \frac{\partial v}{\partial x} \right)}_{\text{pressure-strain}} , \qquad (2,3)$$

$$U \frac{\partial k}{\partial x} + V \frac{\partial k}{\partial y} = \underbrace{- \frac{1}{y^j} \frac{\partial}{\partial y} \left[y^j \, \overline{v \left(\frac{u_i u_i}{2} + \frac{p}{\rho} \right)} \right]}_{}$$

$$\underbrace{\phantom{U \frac{\partial k}{\partial x}}}_{\text{convection}} \qquad \underbrace{\phantom{- \frac{1}{y^j} \frac{\partial}{\partial y}}}_{\text{diffusion}}$$

$$\underbrace{- \overline{uv} \frac{\partial U}{\partial y} - (\overline{u^2} - \overline{v^2}) \frac{\partial U}{\partial x}}_{\text{production}}$$

$$\underbrace{- \nu \, \overline{\frac{\partial u_i}{\partial x_j} \frac{\partial u_i}{\partial x_j}}}_{\varepsilon \, \equiv \, \text{dissipation}} . \qquad (2.4)$$

In (2.4), the summation convention applies where repeated indices appear. In order to obtain a closed system, some of the correlations in (2.1) to (2.4) have to be approximated by model assumptions (how

many depends on the model).   Turbulence measurements
are vital to formulate and test the assumptions.   An
additional transport equation for the energy dissipation rate ε (or the "dissipation" length scale
L $\propto \dfrac{k^{3/2}}{\varepsilon}$) is used in most of the recent models (*e.g.*
Hanjalić and Launder [8], Rodi and Spalding [10]).
Such an equation can also be derived in exact form
from the Navier-Stokes equations.   However, it is not
listed here because the correlations that appear in
it are not as yet accessible to measurement.

When only an equation for k but not for $\overline{uv}$ is
solved, one of the following hypotheses is usually
employed:

1)   Kolmogorov [11], Prandtl [12]:

$$- \overline{uv} \; = \; c_\mu \; \frac{k^2}{\varepsilon} \; \frac{\partial U}{\partial y} \; , \qquad\qquad (2.5)$$

2)   Bradshaw *et al.* [13]:

$$- \overline{uv} \; = \; a_1 k. \qquad\qquad (2.6)$$

c_μ is assumed constant in some models (*e.g.*
Kolmogorov [11], Prandtl [12], Glushko [14], Ng and
Spalding [15]) or made a function of suitable parameters in others (*e.g.* Rodi [16]); a_1 is normally
also assumed a constant.   It is one of the tasks of
turbulence measurements to check these hypotheses.

2.2   The Terms in the Exact Equations

To develop our understanding of turbulent flow
it is desirable to measure as many of the quantities
in the exact equations as possible.   For the present
class of flows, equations (2.1), (2.3), and (2.4) are

the most important ones.   The quantities appearing in
these equations are listed in Table 1.*   They are
divided into three groups according to their importance as target values for test predictions. (Classi-
fication according to the reliability with which they
can be measured would lead to the same grouping.)

   1)   Mean flow quantities.   The longitudinal
mean velocity   U,   and the rate of spread deduced
from it, determine the mean-flow field and are therefore of special interest to the engineer.   Also,
because of its interaction with the turbulent motion,
the mean flow should be known in any study of turbulence.   Therefore,   U   is the quantity of prime experimental interest.

   2)   Reynolds stresses.   The shear stress, $\overline{uv}$,
can be calculated from the momentum equation when the
other variables in this equation (mainly U) are known.
However, for a consistency check on the data, direct
measurement of $\overline{uv}$ is certainly desirable.   The
level of turbulence is represented by the kinetic
energy,   k.   To check whether a model describes the
turbulence field correctly, we thus need measurements
of   k.   These measurements also help to determine
some of the empirical constants and functions.

   3)   Other correlations.   The remaining quantities in Table 1 are necessary to complete the k- and
$\overline{uv}$-balances.   The rate of dissipation, ε,   is of

---

* The lateral velocity,   V,   can be eliminated with
the aid of the continuity equation (2.2).   The only
additional quantities appearing in the equations for
the individual normal stresses $(\overline{u^2}, \overline{v^2}, \overline{w^2})$ are

$$\frac{\overline{p}}{\rho}\frac{\partial u}{\partial x}, \quad \frac{\overline{p}}{\rho}\frac{\partial v}{\partial y}, \quad \frac{\overline{p}}{\rho}\frac{\partial w}{\partial z}.$$

Table 1 Quantities whose measurement is desirable.

Group	1 Mean flow		2 Reynolds stresses			3 Other correlations							
Quantity	U	$\overline{uv}$	$\overline{u^2}$	$\overline{v^2}$	$\overline{w^2}$	ε	$\overline{u^2 v}$	$\overline{v^3}$	$\overline{vw^2}$	$\overline{vp}$	$\overline{uv^2}$	$\overline{up}$	$\overline{p\left(\frac{\partial u}{\partial y}+\frac{\partial v}{\partial x}\right)}$
			$\Big\{$ k		$\Big\}$								
Equations	Mom	Mom	Mom	Mom									
in which	k	k	k	k	k	k	k	k	k	k			
it appears	$\overline{uv}$	$\overline{uv}$		$\overline{uv}$							$\overline{uv}$	$\overline{uv}$	$\overline{uv}$

particular interest because it has a decisive influence in many flows.   Also, it is a quantity for
which some turbulence models solve an equation;   and
it is needed to deduce c_μ from experiments (equation
2.5).   Unfortunately, it cannot as yet be measured
reliably.   Until now, attempts to measure the terms
involving pressure fluctuations in laboratory flows
have not been very successful either.

2.3   *Further Quantities*

There are, of course, many more quantities which
give valuable information about the structure of turbulence, *e.g.* the intermittency factor, two-point and
auto correlations.   These data will certainly be of
increasing importance when, perhaps in a few years'
time, attention is shifted to more complex models in
which these parameters appear.   They have not been
included in this review, however.

3.   EXPECTATIONS FROM THEORY

The exact equations presented above are valuable
tools for evaluating the experimental data.   For
example, if the data are perfect they must satisfy
the various conservation equations precisely;   so they
serve as the basis for consistency checks.   Also
(though not of direct usefulness in a calculation
procedure) the equations provide information on the
theoretically expected general behaviour of the flows,
such as the similarity behaviour.   The present section introduces the theoretical laws which experimental data are expected to follow.

3.1 *Momentum Conservation*

The general behaviour of a free flow depends primarily on the overall momentum excess or deficit of the shear flow relative to that of the external stream. Accordingly, we integrate the momentum equation (2.1) over the whole cross-section of the free boundary layer. For symmetrical flows we obtain (y_E = outer boundary of the layer):

$$(2\pi)^j \frac{\partial}{\partial x} \int_0^{y_E} [U(U-U_E) + (\overline{u^2}-\overline{v^2})] \, y^j \, dy$$

$$+ \quad (2\pi)^j \frac{dU_E}{dx} \int_0^{y_E} (U-U_E) \, y^j \, dy = 0. \qquad (3.1)$$

Equation (3.1) implies that the change with x (the mean flow direction) of the excess momentum flux

$$M = (2\pi)^j \rho \int_0^{y_E} [U(U-U_E) + (\overline{u^2}-\overline{v^2})] \, y^j \, dy \qquad (3.2)$$

is equal to the excess mass flux, $(2\pi)^j \rho \int_0^{y_E} (U-U_E) y^j \, dy$ times the free-stream velocity gradient. Consequently, for flow with constant free-stream velocity, the excess momentum flux M is independent of x. This law is important to check the consistency of the mean-velocity measurements (the term $\overline{u^2}-\overline{v^2}$ is normally negligible in (3.2)).

Three groups of flows. The excess momentum flux M can be positive, negative or zero. Accordingly, symmetrical free flows can be divided into the following three groups:

a)   M > 0.   These flows are jets.   The velocity is, in general, larger than U_E.

b)   M < 0.   These flows are wakes or velocity deficit jets with wake character.   The velocity is smaller than U_E.

c)   M = 0.   These are flows behind self-propelled bodies.   They are also called shear-free layers because, some way downstream of the body, the shear stress is zero (the mean velocity is uniform = U_E).

3.2   *Similarity Analysis*

An important feature of free turbulent boundary layers is their tendency to become self-similar after a certain development region.   Mathematically this means that the terms in the equations of Sec. 2 (convection, diffusion *etc.*) have constant ratios so that the various processes are in dynamic equilibrium, each changing downstream at the same rate as all the others.   As many practically important *non*-similar flows can be considered as transitions between self-similar flows which have different dominant physical processes, the limiting self-similar cases are indeed of special interest.   Also, they facilitate greatly the presentation of data and the execution of consistency checks because they are characterized by one rate of spread and by one dimensionless profile of velocity, kinetic energy, *etc.*   This subsection shows which free flows can be self-similar according to the theory;   and it lists the theoretical laws describing these flows.   The findings will serve as basis for the discussion on the experimental data reviewed in Sec. 5.

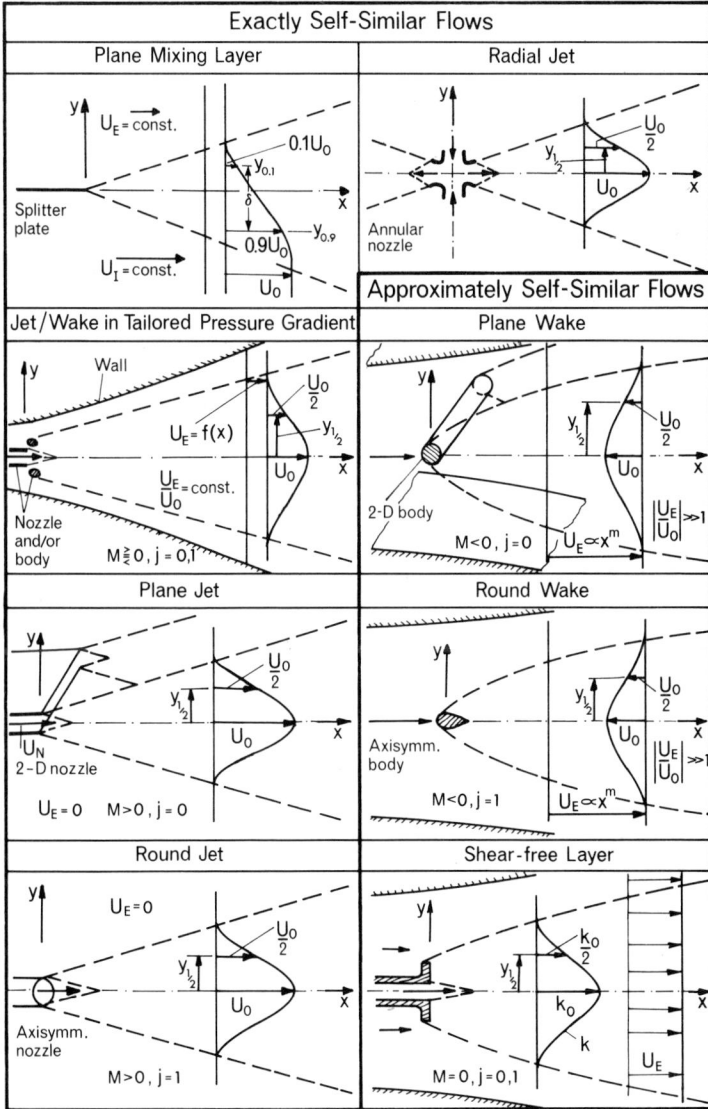

Fig. 2   Potentially self-similar free turbulent
          boundary layers.

a)   Definition of self-similarity

The present definition of self-similarity* is as follows:

A flow is considered self-similar when one velocity and one length scale are sufficient to render its time-averaged quantities dimensionless functions of one geometrical variable only.

The velocity scale chosen here is the maximum excess (or deficit) velocity U_o defined in Fig. 2. For shear-free layers, where U_o is zero, the square root of the maximum kinetic energy, $\sqrt{k_o}$, is used instead. A characteristic width of the flow, δ, is employed as length scale. For the flows considered, δ is defined in Fig. 2 (for symmetrical flows, $\delta = y_{\frac{1}{2}}$).

The above definition of self-similarity implies that the following quantities of importance can be expressed as

$$U = U_E + U_o \; f(\eta)$$

$$V = U_o \; g(\eta)$$

$$k = U_o^2 \; e(\eta)$$

$$\overline{u^2} - \overline{v^2} = U_o^2 \; q(\eta)$$

$$\overline{uv} = U_o^2 \; h(\eta)$$

---

* The term "self-preservation" is often used synonymously in the literature.

$$\overline{v\left(\frac{u_i u_i}{2} + p/\rho\right)} = U_0^3\, \tilde{d}(\eta)$$

$$\varepsilon = \frac{U_0^3}{\delta}\, \tilde{\varepsilon}(\eta) \, ,$$

where
$$\eta = \frac{y}{\delta} \, ,$$

and f, g, e *etc.* are functions of η only. The free-stream velocity U_E is prescribed by constraints on the flow outside the boundary-layer region.

b) *Potentially self-similar flows*

Free turbulent boundary-layers *can* be self-similar only when the governing momentum and energy equations can be written in self-similar form.

Exact self-similarity is possible when (Newman [6])

$$\frac{d\delta}{dx} = \text{const}, \qquad \frac{U_E}{U_0} = \text{const}, \qquad \frac{\delta}{U_0}\frac{dU_0}{dx} = \text{const}.$$

Because of the requirement U_E/U_0 = const, only the following free boundary layers can be exactly self-similar:

(i) *Mixing layers between two uniform-velocity streams*

In this case U_E/U_0 is constant by definition.

(ii) *Jets issuing into stagnant surroundings* (plane, round, radial)

In this case U_E/U_0 = 0 = const.

(iii) *Jets and wakes in an appropriately tailored pressure gradient*

When U_E can be made to vary in such a way that U_E/U_0 remains constant, jets

Table 2   Decay laws of self-similar free turbulent boundary layers.

	FLOW	U_E	$U_O(\sqrt{k_O})$	δ	Re
Exactly self-similar	plane mixing layer	const	const	$\propto x$	$\propto x$
	jet/wake, $\dfrac{U_E}{U_O}$ = const*	$\propto x^{-\frac{j+1}{2}I}$	$\propto x^{-\frac{j+1}{2}I}$	$\propto x$	$\propto x^{1-\frac{j+1}{2}I}$
	plane jet, $U_E = 0$	0	$\propto x^{-\frac{1}{2}}$	$\propto x$	$\propto x^{\frac{1}{2}}$
	round jet, $U_E = 0$	0	$\propto x^{-1}$	$\propto x$	const
	radial jet		$\propto x^{-1}$	$\propto x$	const
Approximately self-similar	jet/wake, $\left|\dfrac{U_E}{U_O}\right| \gg 1$	$\propto x^m$	$\propto x^{\frac{m(j-1)-(j+1)}{j+2}}$	$\propto x^{\frac{1-3m}{j+2}}$	$\propto x^{\frac{m(j-4)-j}{j+2}}$
	plane wake, m = 0	const	$\propto x^{-\frac{1}{2}}$	$\propto x^{\frac{1}{2}}$	const
	round wake, m = 0	const	$\propto x^{-2/3}$	$\propto x^{1/3}$	$\propto x^{-1/3}$
	shear-free layers	$\propto x^m$	$\propto x^{-n/2}$	$\propto x^{1-\frac{n}{2}-m}$	$\propto x^{1-n-m}$

* I is defined as $\quad I = \dfrac{U_E/U_O + I_R}{3/2\ U_E/U_O + I_R}$, where $I_R = \dfrac{\displaystyle\int_0^{\eta_E}(f^2+q)\eta^j d\eta}{\displaystyle\int_0^{\eta_E} f\eta^j d\eta}$

and wakes can be exactly self-similar
for the following ranges of U_E/U_O:

$$- \infty < U_E/U_O \leqslant - 1 \quad \text{and} \quad 0 \leqslant U_E/U_O < + \infty.*$$

Approximate self-similarity when $|U_E/U_O| \gg 1.$
When $|U_E/U_O| \gg 1$, certain terms are negligible in
the momentum and energy equations, and approximate
self-similarity is possible when

$$\frac{U_E}{U_O} \frac{d\delta}{dx} = \text{const}, \qquad \frac{U_E \delta}{U_O^2} \frac{dU_O}{dx} = \text{const}, \qquad \frac{\delta}{U_O} \frac{dU_E}{dx} = \text{const}.$$

These conditions imply that U_E must vary as $U_E \propto x^m$.
Gartshore and Newman [17] give the limits for m as

$$- \frac{j+1}{3} \leqslant m \leqslant - \frac{j}{4-j} .$$

When m is smaller than the lower limit, $|U_E/U_O|$ de-
creases with x and the assumption $|U_E/U_O| \gg 1$ is in-
validated. When m exceeds the upper limit, the
Reynolds number $U_O\delta/\nu$ decreases with x (see Table
2), and the assumption of large Reynolds number is invalid. However, if this happens only very far downstream, there may still exist a limited similarity region. The round wake in zero pressure gradient is the
most important example of a flow of this type.

From the above conditions we infer that approximate
self-similarity is possible in the following flows or
regions of these flows:

   a) Jets and wakes with $|U_E/U_O| \gg 1$.
   The free stream velocity must vary as $U_E \propto x^m$,

---

* When the velocity ratio lies in the range $-1 < U_E/U_O < 0$,
back flow is present and the flow is not a boundary
layer.

and $m \geqslant - \frac{j+1}{3}$. When $m \geqslant - \frac{j}{4-j}$, only a limited similarity region can exist.

b) *Shear-free layers*

Examples are the flow behind a self-propelled body and the turbulence front developing between two equal velocity streams with unequal turbulence energy levels. The maximum kinetic energy decays as $k_o \propto x^{-n}$. The similarity region is limited when $m > 1 - n$.

The above conditions for self-similarity do not restrict the shape of the profiles $f(\eta)$, $h(\eta)$ *etc.*. However, for a given streamwise boundary condition, the profiles will be the same for different upstream conditions.* All the possible boundary conditions are sketched in Fig. 2 which therefore illustrates all the potentially self-similar free flows. Only experiments can tell whether these flows are self-similar in reality.

c) *Decay laws and similarity equations*

The similarity conditions given above, together with the integral momentum equation (3.1), yield the decay laws listed in Table 2. The variation with x of the Reynolds number $U_o \delta / \nu$ is included in the table. To be strictly correct, the power laws of Table 2 should have $(x + x_o)$ as basis, where x_o is the virtual origin of the self-similar flow. This origin depends on the exact conditions of flow generation, such as the velocity profile at the orifice or the shape of the wake-generating body. In all other respects, however, self-similar flows are independent of the initial conditions when the similarity region is unlimited. In

---

* Of course, there is a velocity excess or deficit according to whether a momentum excess or deficit is created upstream.

particular, the constants $d\delta/dx$ (exact self-similarity) and $(U_E/U_o)d\delta/dx$ (approximate self-similarity), and the profile shapes, are independent of
the exact conditions of flow generation.

Tables 3 and 4 list respectively the similarity
forms of the momentum equation (2.1) integrated from
$y = 0$ to y, and of the turbulent kinetic energy
equation (2.3). The former equation can be used to
determine the shear stress from the mean velocity profile, and the latter to evaluate the convection term
in the energy balance. The convection term in the
shear-stress balance is equivalent to that in the
energy balance, with e replaced by h. This study
employs the equations of Tables 3 and 4 to check the
consistency of the data reviewed. These equations
are numbered from (3.3) to (3.19).

3.3 *Non-similar Jets and Wakes*

Jets issuing into a uniformly moving stream
($U_E = $ const.) cannot be self-similar because U_o/U_E
varies with x. The same holds true for wakes for
which $|U_E/U_o|$ is not large. However, in both cases
$|U_E/U_o|$ increases steadily with x until $|U_o|$ is
so small that approximate self-similarity is expected
to prevail. Very far downstream, jets with $U_E > 0$
are therefore expected to behave like self-similar
wakes (but possessing a small velocity excess rather
than deficit). For axisymmetric flow, the Reynolds
number $U_o\delta/\nu$ decreases with x and it is uncertain
whether the development towards a self-similar flow
can occur in this case. Only experiments can provide
the answer.

Jet/wake forgetfulness. For jets, Spalding [18]

Table 3   Shear stress relations for self-similar free turbulent boundary layers.

Flow	Shear stress relation	Eqn. No.
plane mixing layer*	$h = \frac{d\delta}{dx}\left\{\frac{U_E}{U_o}\left(\eta f - \int_o^\eta fd\eta\right) + f\int_o^\eta fd\eta - \int_o^\eta f^2d\eta \right.$ $- (f-1)\left(\frac{U_E}{U_o}\int_o^{\eta_E} fd\eta + \int_o^{\eta_E} f^2d\eta + \int_o^{\eta_E} qd\eta\right)$ $\left. + (\eta + \eta_I)q - \int_o^\eta qd\eta\right\}$	(3.3)
jets and wakes with $\frac{U_E}{U_o} = $ const**	$h = \frac{j+1}{2} \, \text{I} \, \frac{d\delta}{dx}\left\{\frac{1}{\eta^j}\int_o^\eta f\eta^jd\eta\left[\frac{U_E}{U_o}(3 - \frac{2}{I}) + f(\frac{2}{I} - 1)\right]\right.$ $+ \frac{2}{\eta^j}\int_o^\eta f^2\eta^jd\eta(1 - \frac{1}{I}) + \frac{U_E/U_o}{j+1}(\frac{2}{I} - 1)\eta f$ $\left. + \frac{2}{\eta^j}\int_o^\eta q\eta^jd\eta(1 - \frac{1}{I}) + \frac{2}{(j+1)I}\eta q\right\}$	(3.4)
plane jet $U_E = 0$	$h = \frac{d\delta}{dx}\left(\frac{1}{2} f\int_o^\eta fd\eta + \eta q\right)$	(3.5)
round jet $U_E = 0$	$h = \frac{d\delta}{dx}\left(\frac{f}{\eta}\int_o^\eta f\eta d\eta + \eta q\right)$	(3.6)

Table 3 continued.

Flow	Shear stress relation	Eqn. No.
radial jet	$h = \dfrac{d\delta}{dx}\left(f \displaystyle\int_{o}^{\eta} f d\eta + \eta q \right)$	(3.7)
jets/wakes $\|U_E/U_o\| \gg 1,$ $U_E \propto x^m$	$h = \dfrac{U_E}{U_o}\dfrac{d\delta}{dx}\left(1 + \dfrac{j+2}{j+1}\dfrac{m}{1-3m} \right)\eta f$	(3.8)
plane wake $U_E=$const	$h = \dfrac{U_E}{U_o}\dfrac{d\delta}{dx}\,\eta f$	(3.9)
round wake $U_E=$const	$h = \dfrac{U_E}{U_o}\dfrac{d\delta}{dx}\,\eta f$	(3.10)

* For mixing layers, $\eta = y/\delta - \eta_I$, where $\eta_I = y_I/\delta$ and y_I is the position where $U = U_I$. According to Rodi [16]

$$\eta_I = -\frac{1-\lambda^2}{1+\lambda^2}\left(\frac{2\lambda^2}{1-\lambda^2}\int_{o}^{\eta_E} f d\eta + \int_{o}^{\eta_E} f^2 d\eta + \int_{o}^{\eta_E} q d\eta \right)$$

where $\lambda = U_E/U_I$.

** I is defined in Table 2.

Table 4   Turbulent kinetic energy equations for self-similar free boundary layers.

Flow	Kinetic energy equation					Eqn.
	Convection	Diffusion	Shear Stress production	Normal stress production	Diss.	
mixing layer	$-\dfrac{d\delta}{dx}\left(\dfrac{U_E}{U_O}\eta - \dfrac{U_E}{U_O}\int_0^{\eta_E} fd\eta - \int_0^{\eta_E} f^2 d\eta - \int_0^{\eta_E} qd\eta + \int_0^{\eta} fd\eta\right)\dfrac{de}{d\eta}$	$=\dfrac{d}{d\eta}\left(\tilde{d}\,\dfrac{de}{d\eta}\right)$	$-h\dfrac{df}{d\eta}$	$+\dfrac{d\delta}{dx}(\eta+\eta_I)q\dfrac{df}{d\eta}$	$-\tilde{\varepsilon}$	(3.11)
jets/wakes with U_E/U_O=const	$-\dfrac{d\delta}{dx}\left\{(j+1)I\left(\dfrac{U_E}{U_O}+f\right)e+(1-\tfrac{1}{2}I)\left(\dfrac{U_E}{U_O}\eta+(j+1)\dfrac{1}{\eta^j}\int_0^{\eta} f\eta^j d\eta\right)\right\}\dfrac{de}{d\eta}$	$=\dfrac{1}{\eta^j}\dfrac{d}{d\eta}\left(\eta^j\tilde{d}\dfrac{de}{d\eta}\right)$	$-h\dfrac{df}{d\eta}$	$+\dfrac{d\delta}{dx}q\left[\dfrac{j+1}{2}I\left(\dfrac{U_E}{U_O}+f\right)+\eta\dfrac{df}{d\eta}\right]$	$-\tilde{\varepsilon}$	(3.12)
plane jet U_E=0	$-\dfrac{d\delta}{dx}\left(fe+\dfrac{1}{2}\int_0^{\eta} fd\eta\right)\dfrac{de}{d\eta}$	$=\dfrac{d}{d\eta}\left(\tilde{d}\dfrac{de}{d\eta}\right)$	$-h\dfrac{df}{d\eta}$	$+\dfrac{d\delta}{dx}q\left(\tfrac{1}{2}f+\eta\dfrac{df}{d\eta}\right)$	$-\tilde{\varepsilon}$	(3.13)
round jet U_E=0	$-\dfrac{d\delta}{dx}\left(2fe+\dfrac{1}{\eta}\int_0^{\eta} f\eta d\eta\right)\dfrac{de}{d\eta}$	$=\dfrac{1}{\eta}\dfrac{d}{d\eta}\left(\eta\tilde{d}\dfrac{de}{d\eta}\right)$	$-h\dfrac{df}{d\eta}$	$+\dfrac{d\delta}{dx}q\left(f+\eta\dfrac{df}{d\eta}\right)$	$-\tilde{\varepsilon}$	(3.14)

Flow type	Equation	No.
radial jet	$-\frac{d\delta}{dx}\left(2fe + \int_0^\eta f\,d\eta\,\frac{de}{d\eta}\right) = \frac{d}{d\eta}\left(\tilde{d}\,\frac{de}{d\eta}\right) - h\frac{df}{d\eta} + \frac{d\delta}{dx}\,q\left(f + \eta\frac{df}{d\eta}\right) - \frac{2}{3}\tilde{\varepsilon}$	(3.15)
jets/wakes $\|U_E/U_O\| \gg 1$, $U_E \propto x^m$	$-\frac{U_E}{U_O}\frac{d\delta}{dx}\left\{2\,\frac{(j+1)-m(j-1)}{1-3m}\,e + \left(1 + \frac{j+2}{j+1}\frac{m}{1-3m}\right)\eta\frac{de}{d\eta}\right\} = \frac{1}{\eta^j}\frac{d}{d\eta}\left(\eta^j\tilde{d}\frac{de}{d\eta}\right) - h\frac{df}{d\eta} - \frac{U_E}{U_O}\frac{d\delta}{dx}\frac{m(j+2)}{1-3m}\,q - \frac{2}{3}\tilde{\varepsilon}$	(3.16)
plane wake $U_E=$const	$-\frac{U_E}{U_O}\frac{d\delta}{dx}\left(2e + \eta\frac{de}{d\eta}\right) = \frac{d}{d\eta}\left(\tilde{d}\frac{de}{d\eta}\right) - h\frac{df}{d\eta} - \frac{2}{3}\tilde{\varepsilon}$	(3.17)
round wake $U_E=$const	$-\frac{U_E}{U_O}\frac{d\delta}{dx}\left(4e + \eta\frac{de}{d\eta}\right) = \frac{1}{\eta}\frac{d}{d\eta}\left(\eta\tilde{d}\frac{de}{d\eta}\right) - h\frac{df}{d\eta} - \frac{2}{3}\tilde{\varepsilon}$	(3.18)
shear-free layers	$\frac{U_E\delta}{k_O^{3/2}}\frac{dk_O}{dx}\,e - \left(\frac{U_E}{k_O^{1/2}}\frac{d\delta}{dx} + \frac{1}{j+1}\frac{\delta}{k_O^{1/2}}\frac{dU_E}{dx}\right)\eta\frac{de}{d\eta} = \frac{1}{\eta^j}\frac{d}{d\eta}\left(\eta^j\tilde{d}\frac{de}{d\eta}\right) - \frac{\delta}{k_O^{1/2}}\frac{dU_E}{dx}\,q - \frac{2}{3}\tilde{\varepsilon}$	(3.19)

and Bradbury and Riley [19] deduced from dimensional
arguments that, some way downstream from the flow origin, the flow depends only on the excess momentum M*
and not on the nozzle geometry or the ratio of exit to
free stream velocity.   Analogous to this, wake flows
should depend on the deficit momentum only and not on
the geometry of the wake-generating body.   Accordingly,
U_O and δ from different experiments should fall on
one line when plotted versus x/θ, where θ is the
momentum thickness

$$\theta = \left(\frac{M}{\rho U_E^2} \right)^{\frac{1}{j+1}} .$$

This hypothesis will be checked in Secs 5 and 6 with
reference to experimental data.

4.   CRITERIA FOR THE DATA ASSESSMENT

    The present review attempts to select the most
reliable of the many data reported in the literature.
We therefore list below the criteria on which this
selection is based.

Criterion 1:   The actual experimental flow must corres-
                pone closely to the ideal one.

    This criterion is of particular relevance when
the data are to be used for comparison with predictions;
it is also important for consistency checks because
these are based on equations which are valid for certain
flows only.

---

* M is defined by equation (3.2).

We are concerned here with two-dimensional (including axisymmetric) flows only; so the experimental
flow situations should have been closely two-dimensional.   When we consider self-similar flows, the
measurements should have been taken at cross sections
where similarity of both mean and fluctuating quantities prevails.   Lastly, the experimental free-stream
turbulence should have been as low as possible because
its presence complicates further the interpretation of
the data and their comparison with predictions.

Criterion 2:   The instruments and evaluation methods must
            have been adequate.

The data reviewed in this paper were measured with
impact tubes and hot-wire anemometers.   When the turbulence intensity is low, say below 30%, and the probe
is small, the impact tube is a reliable tool for
measuring the mean velocity in free turbulent boundary
layers.   The hot-wire anemometer is the only fully
developed and widely tested instrument for measuring
fluctuating quantities.   It is prone to many errors,
stemming from wire imperfections, calibration instabilities, inadequate electronic equipment, and limitations of the evaluation methods, particularly at high
turbulence levels.   Examination of the equipment and
the evaluation methods employed gives some idea about
the reliability of the data; but a final judgement can
be based on consistency checks only.   In general, the
reliability of the data decreases with increasing turbulence level.

Criterion 3:   The data must be consistent.

The data must pass three consistency tests:

a) The overall momentum must be conserved. This
    is checked by means of the integral momentum
    equation (3.1).

b) The directly measured shear-stress must agree
    with $\overline{uv}$ as determined from the momentum
    equation.   This is checked for self-similar
    flows by means of the equations given in
    Table 3.

c) The diffusion of kinetic energy must integrate
    to zero across the layer (this requirement may
    be seen from equation (2.4)).

5.   REVIEW OF DATA FOR SELF-SIMILAR FLOWS

In this section, experimental data are reviewed
critically for the potentially self-similar flows
sketched in Fig. 2.   The existence of, and the approach
towards, self-similarity are discussed, but the majority
of the results presented are restricted to the self-
similar region.   Data on non-similar jets issuing into
a uniformly moving stream are reviewed in Sec. 6.
Attention is focussed on the quantities discussed in
Sec. 2, in particular on U,   k, $\overline{uv}$ and on the terms
in the energy and shear-stress balances.

5.1   *Plane Mixing Layer*

Flow realization.   The ideal mixing layer is
formed by the mixing between two semi-infinite streams
(see Fig. 2).   This ideal flow can be only approximated
by experiments.   The best approximations are the mixing
between two-dimensional half-jets, and the initial regions of plane and round jets.   However, in the latter

Table 3 Mixing layers with $U_E = 0$.

Flow situation	Experimenter	Half jet dimension	Range	max Re_x	$\overline{u^2}^{\frac12}/U$ at exit	$\frac{d(y_{.1}-y_{.9})}{dx}$	σ	meas. $\frac{\overline{uv}_m}{U_o^2}$	calc. $\frac{\overline{uv}_m}{U_o^2}$	k_m/U_o^2	Remarks
	Reichardt [20]	32×32 cm	<160 cm	3.2×10^6	?	.131 / .15	13.5 / 11.8	---	---	---	most runs / one run
Two-dimensional half-jet with 3 solid boundaries	Liepmann and Laufer [21]	152×19 cm	< 90 cm	1.1×10^6	<1%	.16	11.	.008	.0106	.021	$\overline{w^2}$ was guessed as $\overline{w^2}=\overline{v^2}$
	Wygnanski and Fiedler [22]	51×18 cm	< 58 cm	4.65×10^5	<0.1%	.2	9.	.0091	.0138	.035	wall at x = 0 and trip wire used
	Patel [23]	76×43 cm	<100 cm	1.8×10^6	0.5%	.165	10.7	.0103	.0112	.0275	
	Castro [43]	12.7×76 cm	< 52 cm	10^6	0.12%	.148	11.8	.0083	.0098	.028	
Initial region of a plane jet	Albertson et al. [24]		$1 < \frac{x}{D} < 4$	?	?	.155	11.4				data taken from Abramovich
	Mills [25]	5×15 cm	$2 \leqslant \frac{x}{D} \leqslant 4$	1.8×10^5	?	.178	10.4				
	Sunyach and Mathieu [26]	4×48 cm	$1.5 \leqslant \frac{x}{D} \leqslant 3$	1.6×10^5	0.25%	.18	9.9	.0137	.0125	.05	
Initial region of a round jet	Maydew and Reed [27]		$.5 \leqslant \frac{x}{D} < 3.9$	?	?	.16	11.				exit Mach number = 0.7
	Bradshaw et al. [28]	D = 5 cm	x/D = 2, $2.3 < \frac{x}{r} < 7$	7×10^5	$\cong 0$ 1% at x/D =2	.165	10.7	.010	.011	.027	
	Sami et al. [29]	D = 30 cm	x/D = 3, $3 < \frac{x}{r} < 15$	6.6×10^5	<.5%	.163	10.8	.0109	.0108		

Table 6 Mixing layer between two moving streams.

Flow geometry	Experimenter	Dimensions of nozzles	Range	max Re_x	$\sqrt{\overline{u^2}}/U$ at exit	$\frac{U_E}{U_I}$	$\frac{d(y_{.1}-y_{.9})}{dx}$	σ	meas. $\frac{\overline{uv}_m}{U_o^2}$	calc. $\frac{\overline{uv}_m}{U_o^2}$	k_m/U_o^2	Remarks
Mixing of two free plane jets	Zhestkov *et al.* see Abram. [3]	4×12.5 cm 4×12.5 cm	<10cm	$7.1×10^5$	?	0 to .64						only ratios σ/σ_o are given
	Miles and Shih [30]	not reported	<105cm	$2.6×10^6$	3%	.14 .22 .36 .46 .55 .65 .76 .83		11.6 13.2 19.8 24.3 34.4 34.8 47.0 49.0				σ's taken from Brown and Roshko [34]
	Mills [25]	5×15 cm 5×15 cm	≤20cm	$1.8×10^5$	?	0 .3 .6	.178 .098 .07	10.4 18.3 25.0				
	Seban and Back [31]	.64×30cm 11.5×30cm	<6cm	$1.1×10^5$	?	.2 .36 .6		14.7 16.7 27.9				influence of walls noticeable
	Sabin [32]	not reported	50cm <x<	$1.8×10^5$	?	.35 .46 .6		24.5 31.6				water tunnel

	Size	x / length	Re	Turbulence							Remarks
Spencer [33]	19×38cm 19×38cm	<127cm	2.6×10⁶	.1%	.15 .22 .305 .54 .61 .76	.078 .0328	13.3 15.7 20.4 41.4 50.4 77.0	.0132 .011	.011 .0104	.035 .032	Detailed information only for $\frac{U_E}{U_I}$ = .3, .6
Brown and Roshko [34]	2.5×10cm 2.5×10cm	≤10cm	5×10⁵	.1% to .5%	.143 .378	.123 .084	12.8 18.7				Nitrogen at 7 atm.
Yule [35]	14×50cm 14×50cm	<91cm	1.1×10⁶	≈1%	.30 .61	.095 .046	19 36	– .014	.014 .015	.0346 .0485	Influence of duct boundary layers, particularly for $U_E/U_I = .3$
Gartshore and Pui [37]	45×69cm 45×69cm	50cm <x< 150cm	7×10⁵	≈1% at x=50cm	.65 .75 .81	.042 .0297 .0217	43 61 89	.016 .0143 –	.0165 .0165 .0165	– .0514 –	Honeycombe preceding test section produced free-stream turbulence
Watt [36]					.51	.044	41	.01075	.01085	.0327	Data taken from [7], details were not available to author.

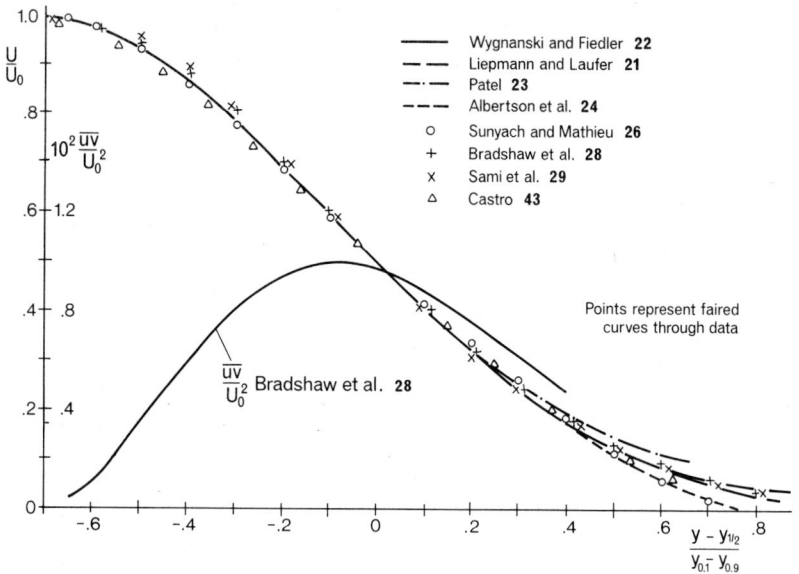

Fig. 3   Velocity and shear-stress profiles for self-
           similar mixing layers with $U_E = 0$.

case, the extent of the mixing layer is rather limited.
Tables 5 and 6 list details of the experimental conditions of the reviewed investigations.   Table 5
covers the mixing layer with one stream at rest ($U_E=0$),
and Table 6 the ones between two moving streams.

   Self-similarity.   From the similarity analysis
of Sec. 3.2 we expect that, downstream of a certain
development region, mixing layers are self-similar for
any velocity ratio U_E/U_I.   This expection is confirmed by all the experimental data on mean and (when
available) turbulence quantities.

   Velocity profiles.   The various experimenters'
similarity profiles for U are shown in Fig. 3 for
$U_E/U_I = 0$ and in Fig. 4 for finite values of U_E/U_I.

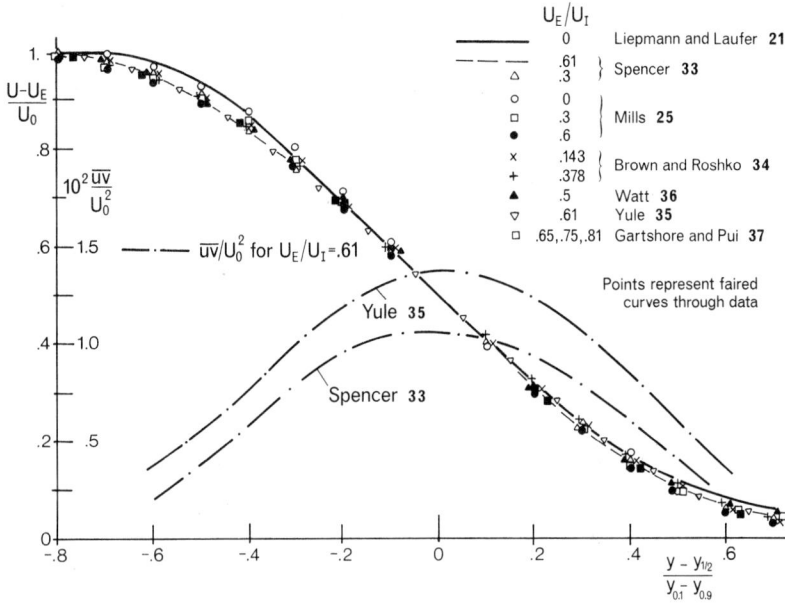

Fig. 4   Velocity and shear-stress profiles for self-similar mixing layers with $U_E/U_I > 0$.

The agreement is good; the somewhat larger scatter in Fig. 3 at the zero velocity edge indicates the difficulties in measuring accurately in this region. However, the velocity profiles agree well only when y is non-dimensionalized by a characteristic flow width, here $\delta = (y_{.1} - y_{.9})$.* When the data are plotted versus y/x, as is often done in the literature, the agreement is poor; for there is little consensus about the rate of spread, $d\delta/dx$.

    Rate of spread. The rates of spread of various experimenters are listed in Tables 5 and 6. The spreading parameter σ is also listed because it appears frequently in the literature. We define σ in such

* $y_{.9}$ and $y_{.1}$ are defined in Fig. 2.

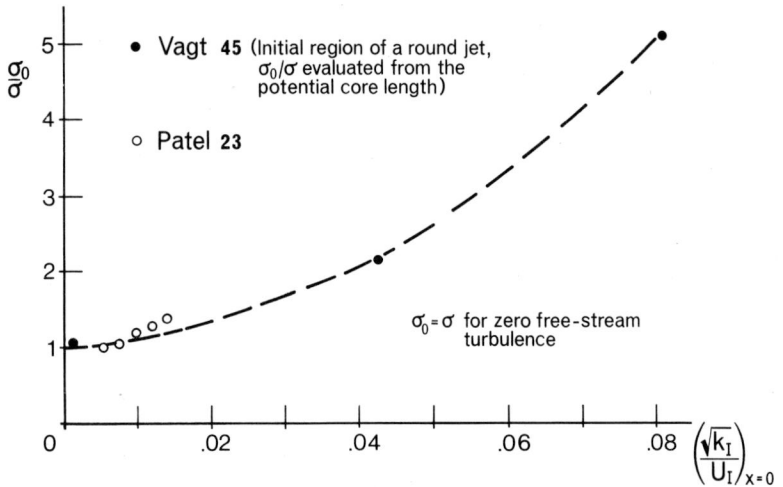

Fig. 5   Influence of free-stream turbulence on the
          rate of spread of a mixing layer with $U_E = 0$.

a way that the use of $\eta = \sigma y/x$ as abscissa brings
the velocity profiles closest to the error function.

Tables 5 and 6 show rather poor agreement about
the rate of spread.   This may be due to the following
three reasons.   Firstly, the distance to attain full
development of the mixing layer is about a thousand
times the momentum deficit thickness of the boundary
layer at the separation point (Bradshaw [39]).   The
data obtained at low values of $Re_x = U_I x/\nu$ are therefore likely to depend on the initial conditions, even
though the velocity profile seems to be developed.
Secondly, the mixing layer seems to be particularly
sensitive to the flow field outside the layer, possibly
because it is an asymmetric flow.   Thus, the layout
of the room into which the layer spreads and room
draughts may influence the rate of spread (Reichardt
[20]).   Thirdly, the influence of free-stream turbu-

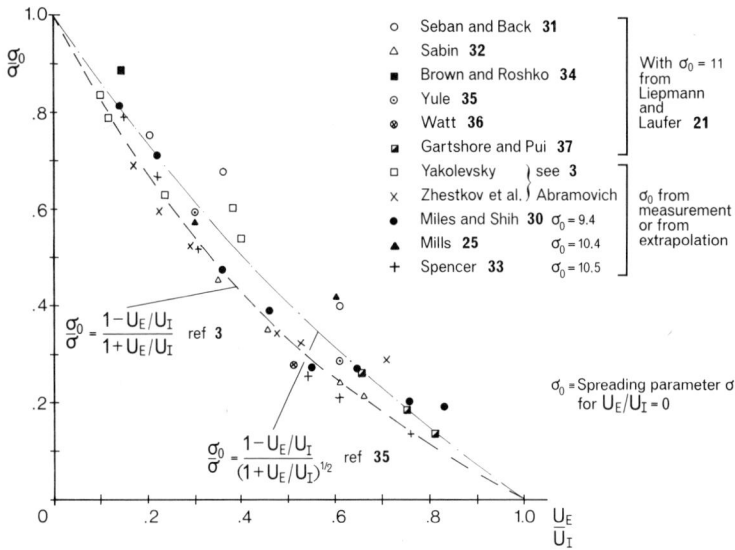

Fig. 6   Influence of the velocity ratio U_E/U_I on the rate of spread of mixing layers.

lence appears to be quite dramatic, as is evidenced in Fig. 5 for $U_E/U_I = 0$.

When the extreme values are discarded in Table 5, $d\delta/dx = .16$ emerges as a fairly well supported value for $U_E/U_I = 0$ and $Re_x > 7 \times 10^5$.

The influence of U_E/U_I on the rate of spread is illustrated in Fig. 6. Similar figures reported in the literature were obtained by taking σ_0, the spreading parameter at $U_E/U_I = 0$, from Liepmann and Laufer [21]. This resulted in considerable scatter. Because σ_0 can vary markedly from one experimental set-up to another (see Table 5), it seems more reasonable to obtain σ_0 by extrapolating each experimenter's results for σ, when enough points are available. This has

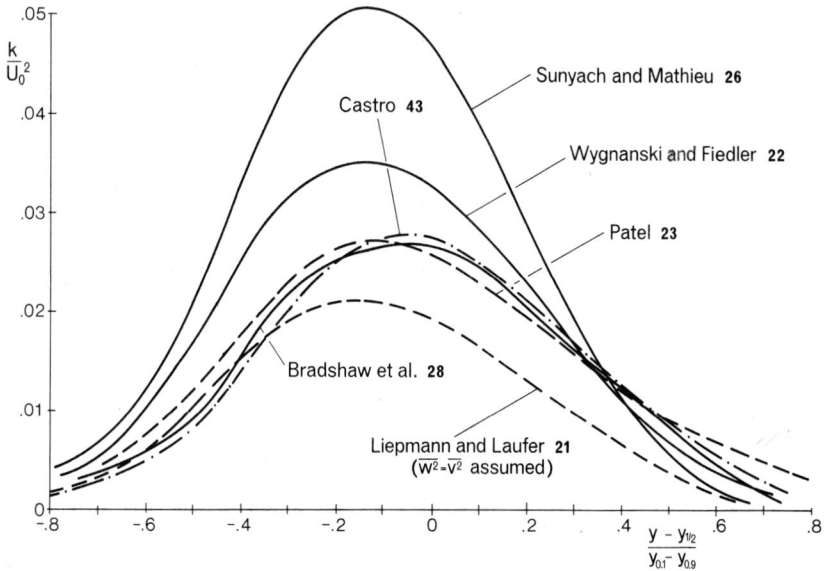

Fig. 7   k-profiles for self-similar mixing layers
          with $U_E = 0$.

been done in Fig. 6;   the scatter was thereby reduced
considerably.   We expect the influence of free-stream
turbulence to grow with increasing U_E/U_I.   As in the
case of $U_E/U_I = 0$,   its effect is to decrease σ.*
The lower points in Fig. 6 should therefore be a better
representation of the ideal flow having no turbulence
in the two free streams.   Abramovich [3] and Yule [35]
have proposed empirical relations for the dependence
of σ_0/σ upon U_E/U_I.   These relations are also
shown in Fig. 6.   Abramovich's relation appears to fit
better the data with low free-stream turbulence and
and Yule's the data with higher free-stream turbulence.

---

* This can be seen by comparing the results of Spencer
[33] and Gartshore and Pui [37].   Figure 8 shows that
Spencer's free-stream turbulence was much lower than
Gartshore and Pui's.

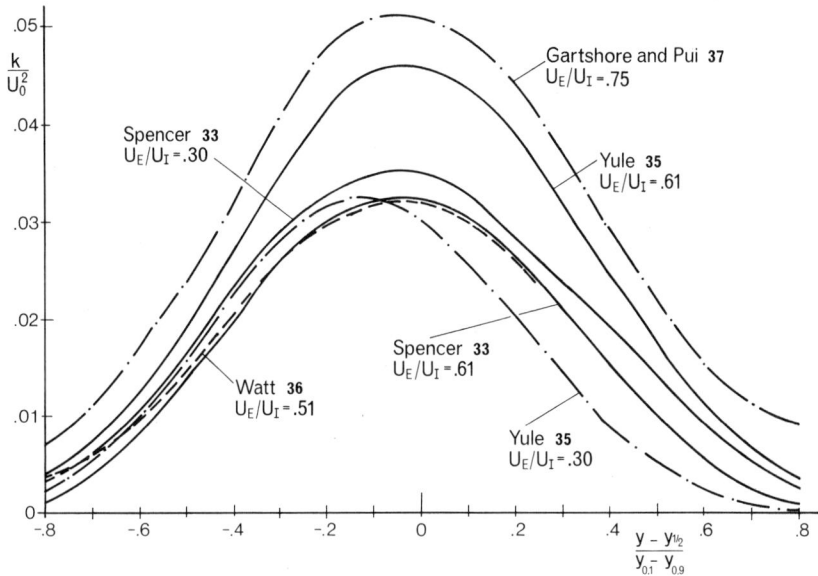

Fig. 8   k-profiles for self-similar mixing layers
with $U_E/U_I > 0$.

Profiles of k.   The similarity profiles of   k
as measured by various experimenters are shown in Fig.
7 for $U_E/U_I = 0$ and in Fig. 8 for $U_E/U_I > 0$. Except
for the shape of the profile, there is no agreement
among measurements for the same value of U_E/U_I.
Liepmann and Laufer's [21] results are probably too
low because of inadequate electronics;   and Sunyach
and Mathieu's [26] are too high because their Re_x
was rather low;   for Laurence [40] and Lassiter [41]
found that $\overline{u^2}$ decreases markedly as Re_x increases.

Consistency check.   To allow further assessment
of the turbulence measurements, Tables 5 and 6 compare
the measured $\overline{uv}_{max}$ with $\overline{uv}_{max}$ as calculated from
the mean velocity via equation (3.3), in which the nor-
mal-stress terms were neglected.

It might be mentioned here that there is considerable confusion in the literature as to how the shear
stress should be calculated from the mean velocity profile.   Normally, an arbitrary assumption concerning
the lateral velocity V is made which is in fact un-
necessary;   for equation (3.3) shows that $\overline{uv}$ depends
only on the rate of spread and on the similarity profile of U when the normal-stress terms are neglected.

Liepmann and Laufer and Wygnanski and Fiedler
claim agreement between their measured and calculated
shear stress.   The author has re-evaluated their data
and found unsatisfactory agreement (Table 5).   For
the mixing layer with U_E/U_I = 0, Bradshaw *et al.'s*
[28] and Patel's [23] results appear to be the most
consistent ones.   They agree remarkably well considering the differences in the geometry of their appar-
atuses (see Table 5).   They are supported by Sami *et
al.'s* [29] data which are also consistent.   Unfortunately, these authors report only the longitudinal
energy component.

The results of Spencer [33] and Yule [35] for
U_E/U_I = .61, of Watt [36] for U_E/U_I = .51, and of
Gartshore and Pui [37] for U_E/U_I = .65 appear all to
be fairly consistent.   The considerable differences
between them are probably due to the influence of free-
stream turbulence.   These differences will be discussed later.

Profiles of $\overline{uv}$, selected for their consistency,
are presented in Fig. 3 for U_E = 0 and in Fig. 4
for U_E/U_I = .61.

Energy balance.   Bradshaw and Ferriss's [38] energy balance is shown in Fig. 9.   It is consistent
because the diffusion integrates to zero across the

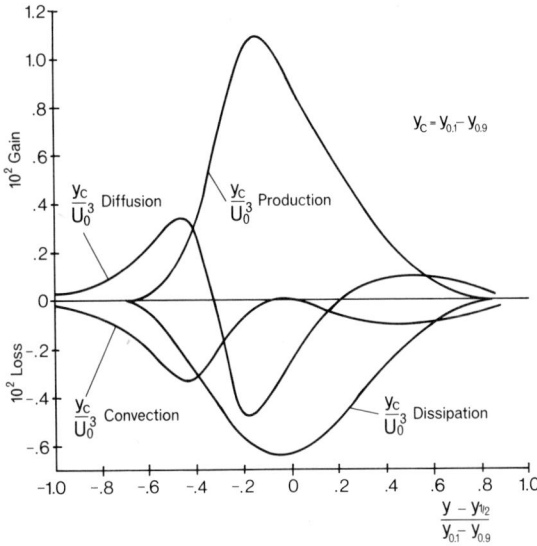

Fig. 9   Energy balance for the self-similar mixing
            layer with $U_E = 0$.   Data of Bradshaw and
            Ferriss [38].

layer.   Sami's [42] balance is in close agreement.
For $U_E/U_I = 0.61$, Spencer's energy balance was modified by multiplying his measured dissipation by a factor of 3.6 to make the diffusion integrate to zero
(the diffusion is determined as the difference of the
other terms).   This improved balance is shown in Fig.
10.
      Shear-stress balance.   Castro [43] measured all
the terms in the shear-stress equation (2.3) except
for those involving pressure fluctuations.   He constructed a $\overline{uv}$-balance by neglecting the pressure diffusion and by determining the pressure-strain term as
the difference of the other terms.   This balance is
shown in Fig. 11.

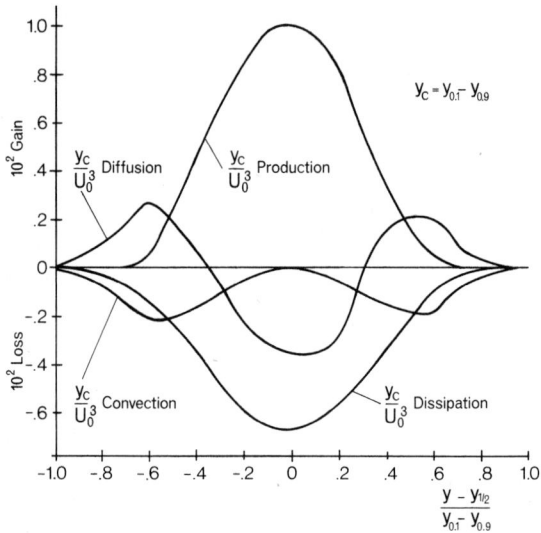

Fig. 10   Energy balance for the self-similar mixing
               layer with U_E/U_I = .61.   Modified data
               of Spencer [33].

Effect of U_E/U_I *on the turbulence structure.*
It is interesting to note that, within the experimental
accuracy, the dimensionless   k-   and $\overline{uv}$-profiles and
the energy balance are the same in Bradshaw *et al.'s*
experiment for U_E = 0   and in Spencer's experiment
for U_E/U_I = .61.   The   k-   and $\overline{uv}$-profiles of Watt
(U_E/U_I = .51) are also in agreement.   This seems to
indicate that the turbulence structure is independent
of the velocity ratio U_E/U_I.   However, on the basis
of their measurements, Yule [35] and Gartshore and Pui
[37] argue that the relative level of   k   and $\overline{uv}$ increases with increasing U_E/U_I.   As there are conflict-
ing results (also for dδ/dx, see Fig. 6) for the same
value of U_E/U_I,   the trend is not entirely clear.
Possibly, the high level of   k   and $\overline{uv}$ (and thus relatively high dδ/dx) in Yule's and Gartshore and

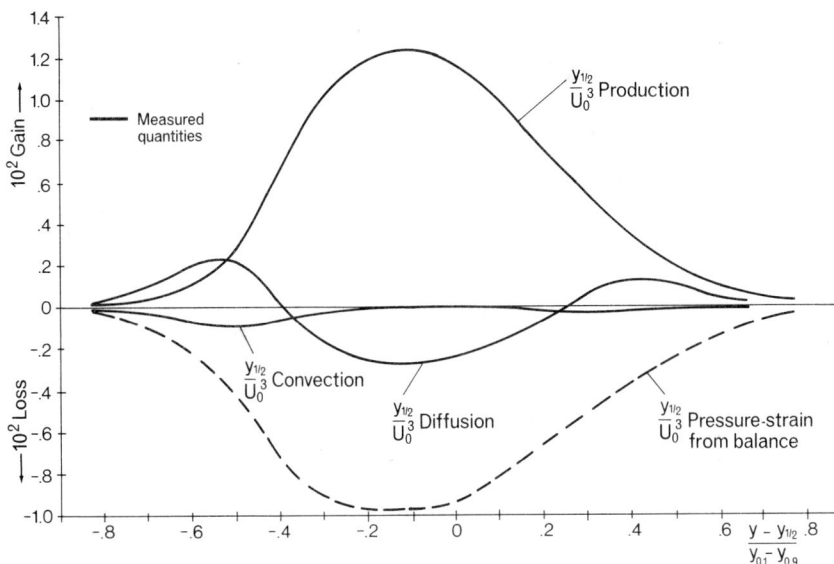

Fig. 11   Shear-stress balance for the self-similar mixing layer with $U_E = 0$.   Data of Castro [43].

Pui's experiments was due to the effect of free-stream turbulence* and not to the effect of U_E/U_I.

5.2 Jets Issuing Into Stagnant Surroundings

Self-similarity.   The experiments confirm our expectation in Sec. 3.2 that jets issuing into stagnant surroundings are self-similar in the far region. The profiles of mean and turbulence quantities become similar (the former ones first), and the centre-line velocity U_o and the jet width $y_{\frac{1}{2}}$ vary as predicted by Table 2.   Except for a shift in the virtual origin

---

* Gartshore and Pui report that the honeycomb preceding their test section generated free-stream turbulence.   From Fig. 8, the turbulence level at the edges of the layer can be seen to be quite high. Yule's flows appear to have been affected by the turbulence generated by the duct boundary layers.

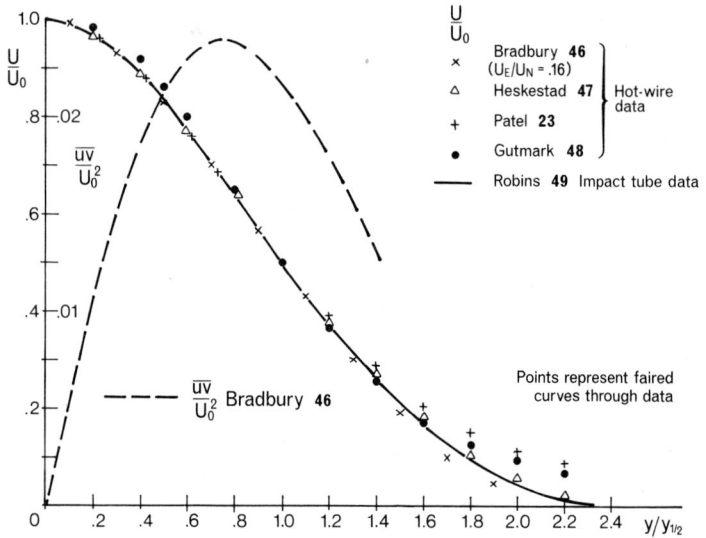

Fig. 12   Velocity and shear-stress profiles in self-
               similar plane jets issuing into still air.

of self-similar flow, the far jet is independent of the
conditions at the nozzle exit.   This is evidenced below
by the general agreement about the rate of spread and
the velocity profile.

Virtual origin.   Flora and Goldschmidt [44] found
that the location of the virtual origin depends, above
all, on the turbulence level at the nozzle exit.   The
same holds true for the potential core length, x_c.
Vagt's [45] results confirm this finding (see Fig. 5).

Table 7 presents the important flow parameters for
the more recent experiments on plane, round and radial
jets.

a)   *Plane jet*

Rate of spread.   From Newman's [6] review on experiments up to 1965, we find an average rate of spread
of $dy_{\frac{1}{2}}/dx \simeq .11$.   Table 7 supports this value.   For

Table 7 Jets issuing into still air.

	Experimenter	Nozzle dimensions	Range (x/D)	Re_D	Start of self-pres. region	$\dfrac{dy_{\frac{1}{2}}}{dx}$	meas. $\overline{uv}_m/U_o$	calc.** $\overline{uv}_m/U_o$	$\dfrac{k_\varepsilon}{U_o^2}$	Remarks
plane jet	Bradbury [46]	D = .95 cm L = 46 cm	14-70	3×10^4	x/D ≈ 30	not constant	.0242	a) 0.024 b) 0.0249	.067	$U_E/U_N = .16$
plane jet	Heskestad [47]	D = 1.25 cm L = 150 cm	47-155	$.47\times10^4$ -3.7×10^4	x/D ≈ 65 but u_t^2/U_o^2 keeps rising	.11	.021	a) 0.0232 b) 0.0277	.07	strong influence of Re_D on $\sqrt{u_t^2}/U_o$
plane jet	Patel [23]	D = .7 cm L = 80 cm	12-152	3.5×10^4	For U and u^2:x/D=30	.103	.021	a) 0.0225 b) 0.0247	.064	
plane jet	Gutmark [48]	D = 1.3 cm L = 50 cm	10-150	3×10^4	x/D ≈ 120	.102	.024	a) 0.0238 b) 0.0287	.077	
plane jet	Robins [49]	aspect ratios 21-138	5-100	1×10^4 -6×10^4	x/D ≈ 60	.103	.02	a) 0.023 b) 0.025	.064	
round jet	Wygnanski and Fiedler [51]	D = 2.6 cm	20-98	10^5	x/D ≈ 70	.086	.0165	a) 0.0168 b) 0.0190	.101	strange change in slope of U_N/U_o versus x/D at x/D=57
round jet	Rodi [16]	D = 1.29 cm	62-75	8.7×10^4	x/D < 62	.086	.0186	a) 0.0168 b) 0.0188	.092	
radial jet	Tuve [52]	annular slot widths D = 2.54 cm D = .635 cm	6-95	$.6\times10^4$ -3×10^4	?	total angle 23°				
radial jet	Heskestad [53]	width of annular slot D = 0.71 cm	20-112	2.5×10^4	only U self-pres. from x/D=46	.111	.0254	a) 0.046 b) 0.0508*	.096 at x/D=65	for x/D > 100, influence of walls noticeable

* Heskestad [53] reports a value of 0.045

** a) normal-stress term neglected in (3.5) to (3.7)
   b) normal-stress term included in (3.5) to (3.7)

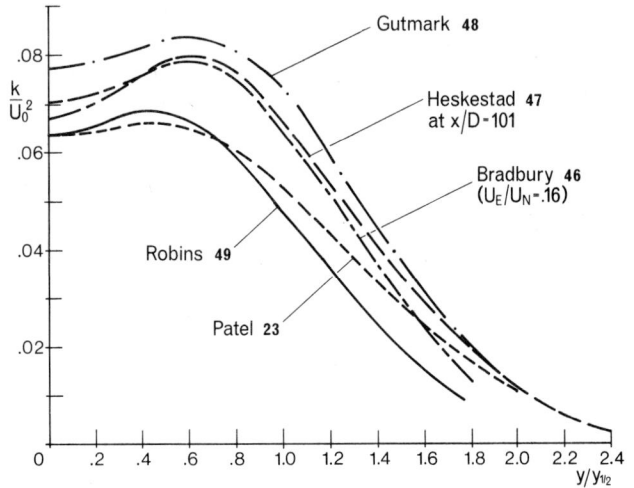

Fig. 13   k-profiles in self-similar plane jets
               issuing into still air.

low turbulence intensities at the exit, the potential
core length is $x_c/D \simeq 6$.

Velocity profile.   Figure 12 shows various experimenters' similarity profiles of U.   Near the
edge, there is considerable disagreement.   Since hot-
wire anemometers are known to measure too high in this
region (see Rodi [16]), the lower values are probably
more accurate.   Robins' [40] curve is recommended as
target profile because Bradbury's [46] profile is probably influenced near the edge by the slow but finite
free stream he used.

Start of self-similarity.   True self-similarity
starts where the turbulence quantities become similar
(the mean velocity does so much earlier).   Table 7
shows that there is no consensus about the position of
this start.   These differences may be attributed, in
part, to the difficulties in maintaining the flow two-

dimensional over long distances. Also, the initial
conditions may influence the turbulence quantities
farther downstream than is commonly thought.

k-profiles. Figure 13 compares the k-profile
of various experimenters. The agreement is rather poor.
Heskestad's [47] results are suspect because his
$\overline{u_\ell^2}/U_o^2$ still rises at x/D = 160, and because he
found a strong influence of the Reynolds number (DU_N/ν).
Gutmark's [48] results lose credibility because he
reports some oddities: his velocity decay changes
abruptly at x/D = 65, and the ratio of dissipation
to production in his energy balance is of the order of
0.2, which is not plausible. Robins [49] found $\overline{v^2}$
and $\overline{w^2}$ to be very much smaller than $\overline{u^2}$ near the
edge, which is contrary to all other measurements.

Consistency checks. For a consistency check,
the measured and calculated values of $\overline{uv}_{max}$ are compared in Table 7. Two calculated values are listed:
one was obtained from the full equation (3.5), and
one with the normal-stress term neglected in (3.5).
It is not surprising that Bradbury's [46] data are the
most consistent ones because his jet issued into a
slow moving stream (U_E/U_N = .16). This stream reduced the relative intensities and thus the measurement errors.

Bradbury's data. Bradbury's jet could not be
exactly self-similar (because U_E > 0), but the de-
partures from self-similarity were insignificant in
the region investigated. Also, reducing U_E/U_N from
0.16 to 0.07 had no effect on the results, nor had
halving the Reynolds number. Despite the fact that
Bradbury is the only one to report the somewhat im-
plausible result that $\overline{v^2} > \overline{u^2}$ at the centre, his tur-

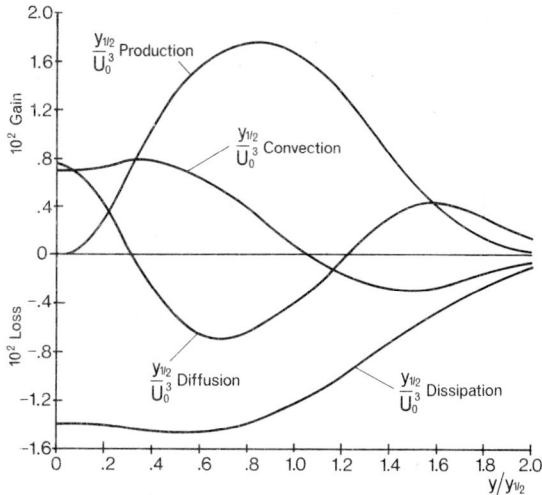

Fig. 14   Energy balance for the self-similar plane
               jet issuing into still air.   Data of
               Bradbury [46].

bulence data are recommended as the most reliable
ones.   His $\overline{uv}$-profile is presented in Fig. 12.

     k- *and* $\overline{uv}$-*balances.*   Bradbury's energy balance
is shown in Fig. 14.   It is consistent.   Robins and
Gutmark measured all the terms necessary to construct
the shear-stress balance except for the terms involving pressure fluctuations.   Unfortunately, the general
consistency of their measurements is not very good.

     Thus, for the construction of the shear-stress
balance, only the diffusion term was taken from Robins,
and the other terms from Bradbury and Patel.   This
shear-stress balance is shown in Fig. 15.

     b)   *Round jet*

     The mean flow field of round jets has been studied
by numerous experimenters, but the turbulence field in
the truly self-similar region only by few.   Since
Gibson's [50] jet was not really symmetrical, only

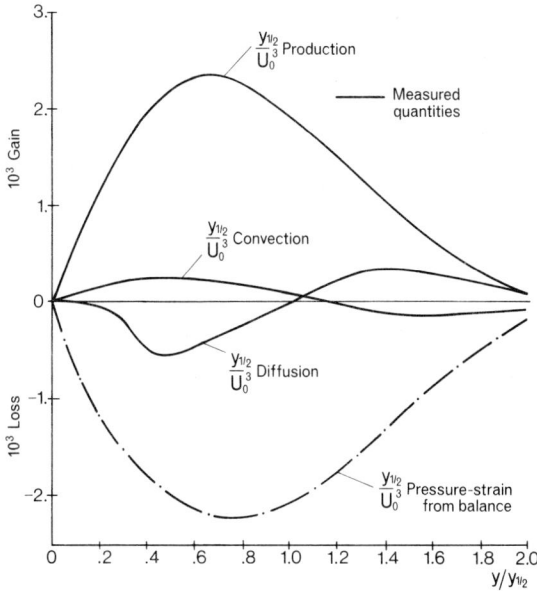

Fig. 15   Shear-stress balance for the self-similar
             plane jet issuing into still air. Origin
             of data: $\overline{v^2}$ from Patel [23], $\overline{uv}/k$, ε,
             $\partial U/\partial y$ from Bradbury [46], $\partial(\overline{uv^2})/\partial y \equiv$
             diffusion from Robins [49].

Wygnanski and Fiedler's [51] and Rodi's [16] measurements can be included.

Rate of spread.   Newman [6] reports a rate of
spread of $dy_{\frac{1}{2}}/dx = .086$ as average over earlier experiments.   The more recent data listed in Table 7
support this value.   For low exit turbulence level,
the potential core length is $x_c/D \simeq 5$ (Vagt [45]).

Profile of U, k *and* $\overline{uv}$.   Rodi's [16] velocity
profile is shown in Fig. 16.   It agrees well with the
earlier measurements.   Wygnanski and Fiedler's and
Rodi's k-profiles are presented in Fig. 17.   Compared
with the situation in other flows, the agreement is
reasonable.   Rodi's results are more consistent, as

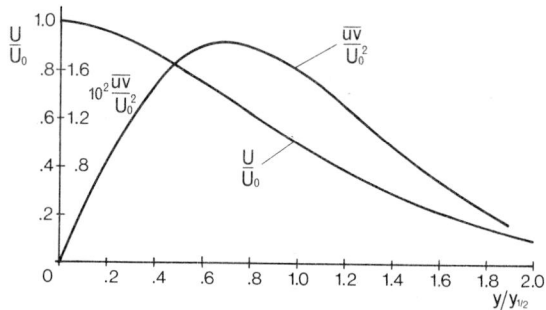

Fig. 16   Velocity and shear-stress profiles in the
               self-similar round jet issuing into still
               air: data of Rodi [16].

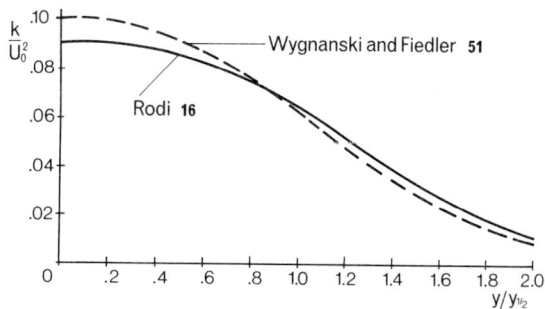

Fig. 17   k-profiles in the self-similar round jet
               issuing into still air.

can be seen from the measured and calculated $\overline{uv}_{max}$-
values of Table 7.   His $\overline{uv}$-profile is shown in Fig.
16.

       k- *and* $\overline{uv}$-*balances.*   Wygnanski and Fiedler's
energy balance has two defects: the normal-stress
production is too large by a factor of two, and the
diffusion does not quite integrate to zero.   An
improved version is shown in Fig. 18.   The shear-
stress balance constructed from Wygnanski and Fiedler's
measurements is shown in Fig. 19.

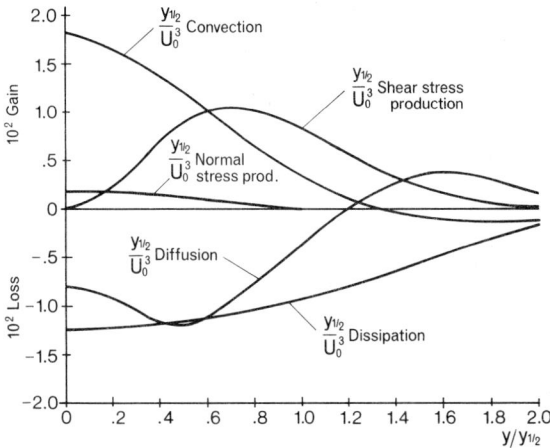

Fig. 18   Energy balance for the self-similar round
            jet issuing into still air: modified data
            of Wygnanski and Fiedler [51].

c) Radial jet

The radial jet emerging from an annular nozzle
has been investigated by two experimenters only. Tuve
[52] measured the mean velocity, and Heskestad [53]
both mean and turbulence quantities. The mean velocity
became self-similar in both experiments. The *velocity
profiles* are shown in Fig. 20. Heskestad's hot-wire
data are probably too high at the edge. He measured
a *rate of spread* of $dy_{\frac{1}{2}}/dx = .11$, which is compatible
with $dy_E/dx = .42$ as measured by Tuve, where y_E is
the total width of the shear layer.

Lack of self-similarity. Heskestad's turbulence
quantities are not self-similar in the region investi-
gated. $\overline{u_{\text{\c}}^2}/U_O^2$ keeps rising, and the intermittency
profile does not even approach a self-similar form.
Beyond $x/D = 90$, $\gamma_{\text{\c}}$ decreases rapidly. Heskestad
notes that the jet flow breaks down at this point be-

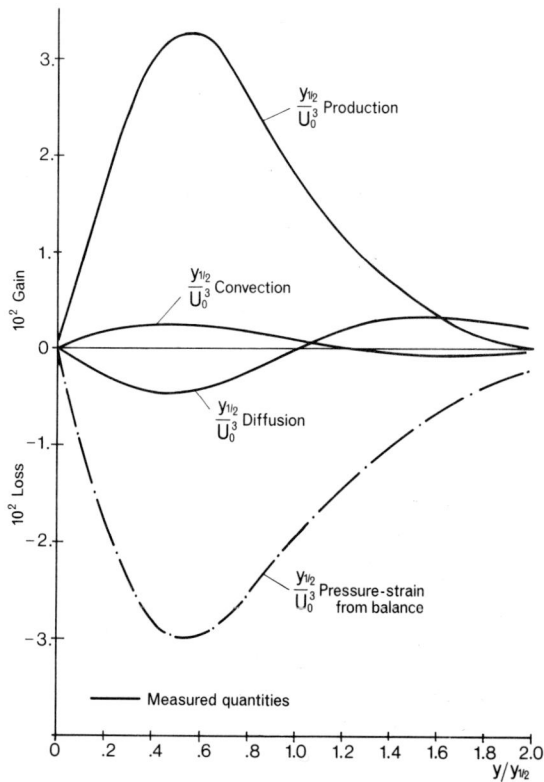

Fig. 19   Shear-stress balance for the self-similar round jet issuing into still air. Constructed from data of Wygnanski and Fiedler [51].

cause of room disturbances.   Owing to this lack of true self-similarity, the turbulence quantities are not presented here.

5.3   *Self-similar Jets and Wakes in Appropriately Tailored Pressure Gradients*

The similarity analysis of Sec. 3.2 revealed that jets and wakes in a streamwise pressure gradient can

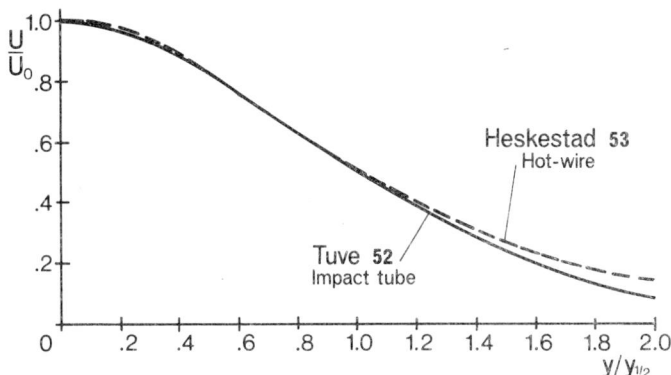

Fig. 20   Velocity profiles in self-similar radial
               jets.

be exactly self-similar when U_E/U_O is kept constant.
Experimental verification of this result may be found
in the work of Fekete [54] for jets and of Gartshore
[55] for wakes.   Fekete investigated jets with U_O/U_E
ranging from .265 to .950 and both his mean and
turbulence data confirm well the similarity behaviour
described in Sec. 3.2.   Gartshore studied two wakes
with U_E/U_O equal to -.192 and -.24 respectively.
The first wake showed quite good similarity behaviour,
the second one did so only to a lesser degree.   Both
Fekete and Gartshore report that their measured shear
stress agrees with $\overline{uv}$ as determined from the momentum
equation.

The variation of the rate of spread, $dy_{\frac{1}{2}}/dx$,
with U_O/U_E is shown in Fig. 21.   Fekete's similarity
profiles for U and $\overline{uv}$ are shown in Fig. 22 for various ratios of U_O/U_E.   The corresponding k-profiles
are presented in Fig. 23.   The profiles for $U_O/U_E \simeq \infty$
(jet in still air) reported earlier are also included
in Figs 22 and 23.   Fekete's velocity profiles are independent of U_O/U_E; and from the profiles of $\overline{uv}$ and
k no clear influence of U_O/U_E can be detected. It

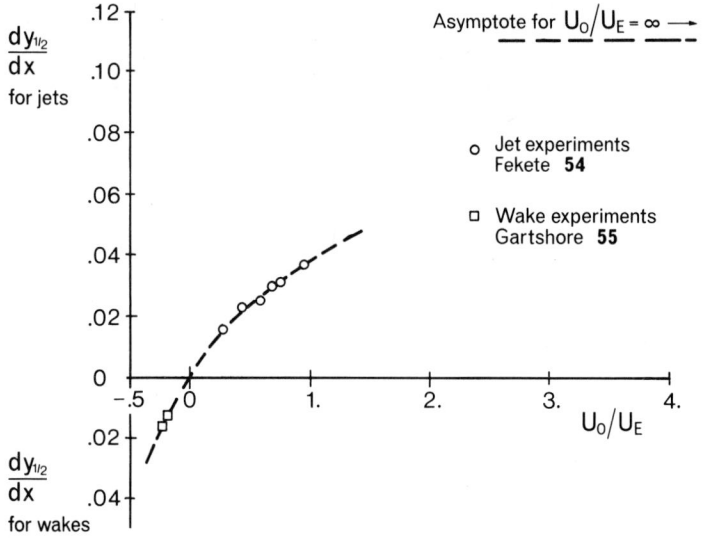

Fig. 21   Influence of the velocity ratio U_O/U_E on the rate of spread of exactly self-similar plane jets and wakes (in appropriately tailored pressure gradients).

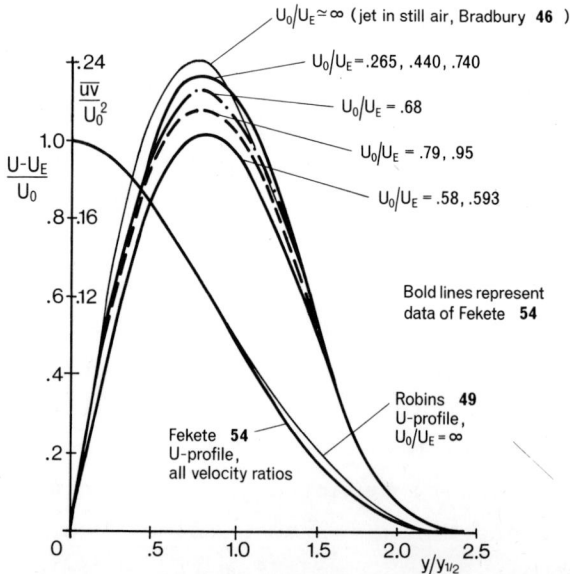

Fig. 22   Velocity and shear-stress profiles in self-similar plane jets in a pressure gradient.

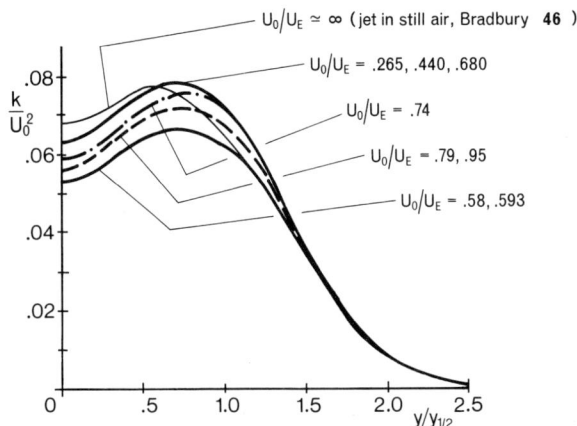

Fig. 23   k-profiles in self-similar plane jets in
             a pressure gradient.

appears therefore that the velocity ratio U_O/U_E has
no significant influence on the turbulence structure
in self-similar plane jets.

5.4   *Wakes in Zero Pressure Gradient*

Conditions for self-similarity.   We are interested
here mainly in the far wake region in which approximate
self-similarity as described in Sec. 3.2 is expected
to prevail.   Such a region can exist only when certain
terms in the momentum equation are negligible.   To
elucidate this condition, we write the total deficit
momentum flux as

$$M \ = \ M_W (1 + \frac{U_O}{U_E} I_R) = \text{const},$$

where

$$M_W \ = \ (2\pi)^j \ \rho \ U_E U_O \ y_{\frac{1}{2}}^{j+1} \int_O^\infty f\eta^j d\eta \ ,$$

and I_R is a shape parameter defined in Table 2.   Experiments have shown that I_R is about   .72   for   j=0

and .54 for j=1.

The similarity analysis for wakes (approximate self-similarity, Sec. 3.2) assumes the term $U_o/U_E I_R$ to be negligible compared with unity. If this term is to contribute less than, say, 3% to the total deficit momentum flux M, then $|U_o/U_E|$ must be smaller than $\simeq.06$. Tables 8 and 9 show that most wake experiments met this condition; the data should therefore evidence the expected similarity behaviour.

Determination of the spreading parameter. With the aid of the integral momentum equation (3.1), the spreading parameter $S = (U_E/U_o)(dy_{\frac{1}{2}}/dx)$ can be expressed as

$$S = \frac{1}{j+2} \frac{d}{dx}\left(\frac{U_E}{U_o} y_{\frac{1}{2}}\right) . \qquad (5.1)$$

For the present review, S was determined from experiments by plotting $U_E/U_o y_{\frac{1}{2}}$ versus x; the slope then yields S by means of (5.1).

S and other important wake parameters are listed in Table 8 for plane wakes and in Table 9 for round wakes.

a) Plane wakes

Self-similarity. Far downstream of the body, plane wakes behave as predicted by the similarity analysis: U_o varies as $x^{-\frac{1}{2}}$ and $y_{\frac{1}{2}}$ as $x^{\frac{1}{2}}$; and similarity of both mean and turbulence quantities prevails. According to Townsend [56], the similarity region starts at $x/D \simeq 500$ in the wake of a cylinder (D = diameter of cylinder). The approach towards similarity of the turbulence quantities is shown in Fig. 24. It is noteworthy that the asymptotic values are approached from below in the wake of a flat plate,

Table 8   Plane wakes in zero pressure gradient.

Body	Experimenter	c_D	Re_D	Re_θ	$\dfrac{x}{D}$	$\dfrac{x}{\theta}$	$\left|\dfrac{U_E}{U_o}\right|$	s	$\dfrac{y_{\frac{1}{2}}}{[\theta(x-x_o)]^{\frac{1}{2}}}$	$\dfrac{\sqrt{\overline{u_\ell^2}}}{U_o}$	meas. $\dfrac{\overline{uv}_m}{U_o^2}$	calc. $\dfrac{\overline{uv}_m}{U_o^2}$	Remarks
circular cylinder	Reichardt [20] Schlichting [57]	1.32	2.38×10^4	1570		60-3400	>.025	.1	.328				similarity from x/D = 400
	Townsend [56,58] **		1360		80-950		.093-.031	.098	.313	.27	.0505	.051	similarity from x/D = 600
	Ermshaus [59]	.94	2500	1170	100-575	200-1230	.095-.044	.089	.303†				
	Ermshaus [59]	1.09	6200	3370	80-240	140-440	.103-.066			.345*	.035*		
	Everitt [60]				120-800	300-1600	.07-.038	.096	.334	.259	.0364	.050	$|U_o y_{\frac{1}{2}}|$ starts to rise beyond x/D = 450
	Alexopoulos and Keffer [61]		5000		40-228		.159-.072	.107	.41†	.3	.033	.055	wake not yet developed
	Uberoi and Freymuth [62]		540 & 4320		25-800		not reported			.305			only $\overline{u^2}/U_E^2$ reported. Townsend's U_o/U_E used to determine $\overline{u^2}/U_o^2$
plate	Ermshaus [59]	1.88	2340	2200	40-612	44-650	.206-.068	.805	.3†				
aerofoil	Ermshaus [59]	.35	3000	525	68-463	400-2600	.066-.029	.098	.349				
	Everitt [60]				50-800	50-800	.18-.047	.172	.44	.259			
flat plate	Chevray and Kovasznay [63]			3300	0-210	0-210	.31-.134	.062*					wake far from being developed
jet in moving stream	Everitt [60]				<314		>.158	.0866*		.224*			velocity excess

* Quantity continues to change with x.

† Exponent of $y_{\frac{1}{2}}$ is not $\frac{1}{2}$.

** The following quantities are not correct in Townsend's [56] book: $\overline{uv}/u^2$ in Fig. 7.10, $u^{2\frac{1}{2}}/U_E$ in Fig. 7.1.

Table 9   Round wakes in zero pressure gradient.

Body	Experimenter	c_D	Re_D	x/D	$\left\|\dfrac{U_0}{U_E}\right\|$	s	S_t	$\dfrac{\sqrt{\overline{u_t^2}}}{U_0}$	meas. $\dfrac{\overline{uv}_{max}}{U_0^2}$	calc. $\dfrac{\overline{uv}_{max}}{U_0^2}$	Remarks
round jet in moving stream	Reichardt [66]	?	$>10^4$	<160	>.053	.105*					velocity excess
spheroid 1:6	Chevray [67]	.06	2.75×10^6	0-18	.28-.091	.105*		.3*	.041*	.054	flow not developed at x/D = 18
cone 1:6	Reichardt and Ermshaus [68]	.68(?)	?	20-60	?	.14					
cone 1:4	Reichardt and Ermshaus [68]	.75(?)	?	10-60	?	.21					
	Ermshaus [59]	.75(?)	2.5×10^5	10-60	.167-.041	.23		.46*	.072*	.12	
cone 1:2	Reichardt and Ermshaus [68]	?	?	30-50	?	.57					
sphere	Uberoi and Freymuth [69]	.4(?)	8600	10-300	.012 at x/D=100	.51	1.	.85	.27	.27	values quoted are for x/D = 100
cone 1:1, disk	Reichardt and Ermshaus [68]	?	?	20-50	?	.86					
disk	Ermshaus [59]	1.1	2.5×10^4	10-60	.078-.017	.93		.98	.5	.56	
disk	Carmody [70]	1.16	7×10^4	≤15	>.055	.8*		.94	.5	.48	flow not fully developed at x/D=15
disk	Cooper and Lutzky [72]	?	3600 / 8300	23-680			.92 / 1.22				mean velocity not reported
disk	Hwang and Baldwin [73]	?	7.7×10^3 -3.1×10^4	4-800			1.21				mean velocity not reported
square plate	Cooper and Lutzky [72]	?	1.45×10^4	80-680	.0153-.0037	.92	1.23	.95		.525	body not axisymmetric

* Values still change with increasing x.

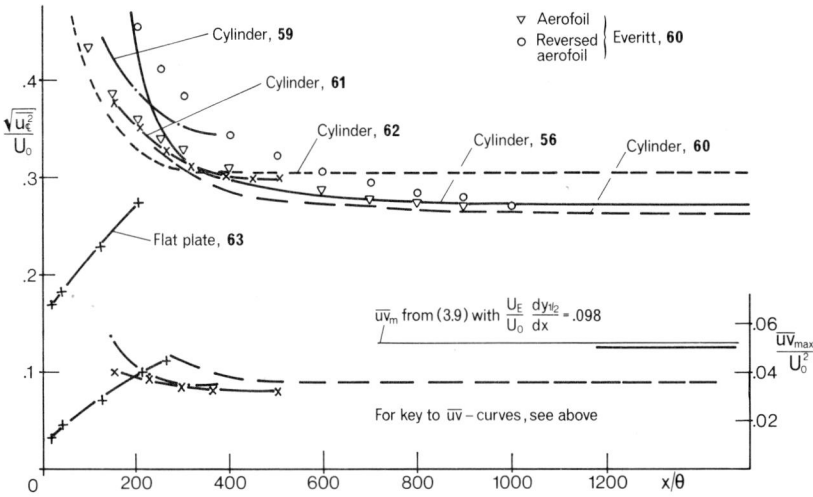

Fig. 24   Relative centre-plane intensity and maximum shear-stress in plane wakes.

while they are approached from above in other wakes.
Starting from a wall boundary layer, the wake has to
build up turbulence energy; in contrast, the vortex
shedding behind blunt bodies generates a surplus of
energy which has to decay before an equilibrium can
exist.

Influence of initial conditions.   We expect that
self-similar far wakes should be independent of the
precise conditions of their generation.   Thus, the
body shape and size, and the free-stream velocity,
should have no direct influence, but only the momentum
thickness θ (see Sec. 3.3).   To test this presumption,
$y_{\frac{1}{2}}/(Dc_D)$ is plotted versus $x/(Dc_D)$ in Fig. 25. θ is
related to the drag coefficient c_D by $\theta = Dc_D/2$.
Only at very large values of $x/(Dc_D)$ is there some
agreement (although primarily between cylinder-wake
data).   Thus, the initial conditions appear to exert

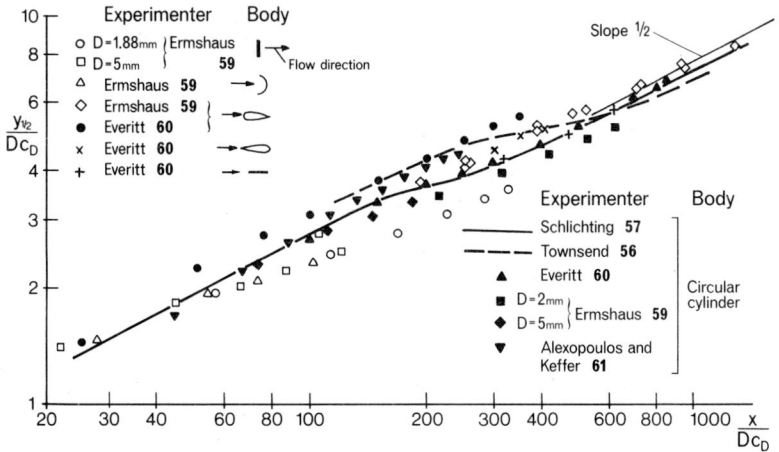

Fig. 25 $y_{\frac{1}{2}}/(Dc_D)$ versus $x/(Dc_D)$ for plane wakes.

their influence very far downstream.

Spreading parameter.   There is, however, some
encouraging agreement about the spreading parameter
S.   When the extreme values in Table 8 are excluded,
Townsend's [56] value of   S = .098   is supported fairly
well by the rest of the data.

The spreading parameter $y_{\frac{1}{2}}/[\theta(x-x_o)]^{\frac{1}{2}}$ is also
listed in Table 8 because it is more popular in the
literature.*   Its calculation requires knowledge about
c_D and the somewhat arbitrary virtual origin x_o.   We
therefore prefer the parameter   S,   which can be determined from U_o/U_E and $y_{\frac{1}{2}}$ alone.

Profiles of U, k *and* $\overline{uv}$.   The velocity profiles
of various experimenters are shown in Fig. 26.   They

---

* The conversion formula is $S = \dfrac{y_{\frac{1}{2}}^2}{\theta(x-x_o)} \displaystyle\int_0^{\infty} f d\eta.$

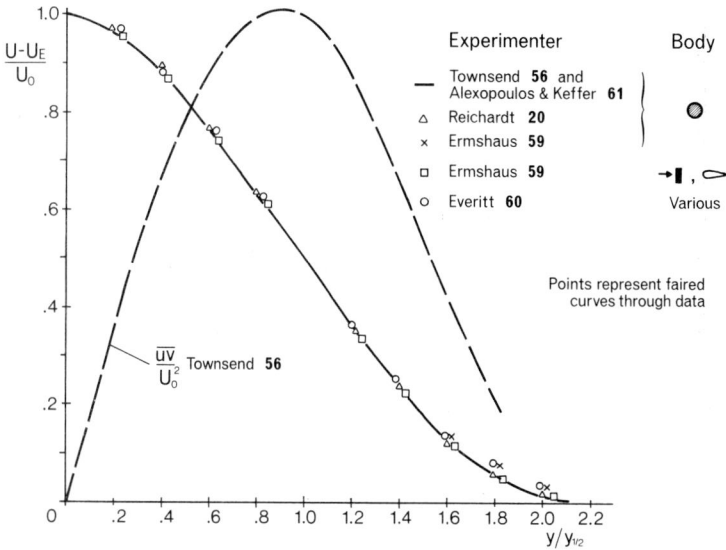

Fig. 26   Velocity and shear-stress profiles in
               self-similar plane wakes.

agree very well.   The k-profiles are compared in Fig.
27.   They are all somewhat suspect for the following
reasons.   Table 8 reveals that Everitt's [60] and
Alexopoulos and Keffer's [61] measured values of
$\overline{uv}_{max}$ are 30% and 40% lower than the ones calculated
by means of equation (3.9).   Their kinetic energy is
likely to be too low also.   Townsend's [58] measurements are consistent.   His $\overline{uv}$-profile is shown in
Fig. 26.   However, the data of Uberoi and Freymuth
[62], of Chevray and Kovasznay [63], and of Thomas
[64], which appear to be of high equality, suggest
that $\overline{u^2}$ is higher than that found by Townsend.

   Energy balance.   Townsend's [58] energy balance
for the far wake is not consistent.   His balance
quoted in the books of Townsend [56] and Hinze [65]
was measured at $x/D = 160$,   which was too close to

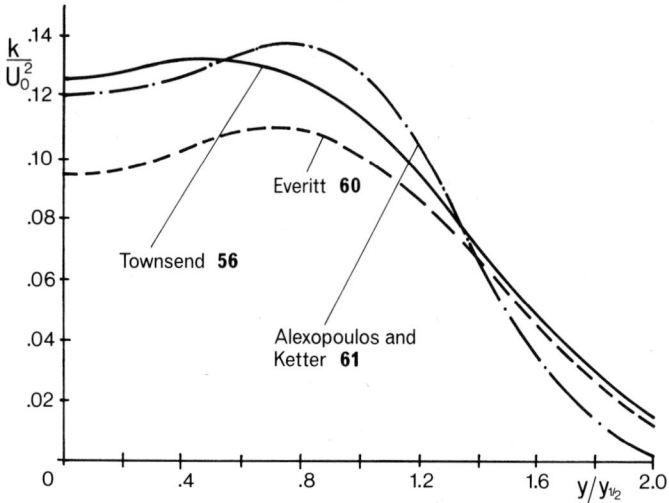

Fig. 27   k-profiles in self-similar plane wakes
             (behind circular cylinders).

the body for the wake to be self-similar.   This balance is therefore not included here.

    b)   *Round wakes*

    Existence of self-similarity.   Round wakes,
together with round jets in a moving stream, are unique
among free shear flows in that their local Reynolds
number ($\equiv U_0 y_{\frac{1}{2}}/\nu$) decreases with x.   As a consequence, only a limited similarity region can exist;
and the question arises whether it exists at all.   The
answer of all the measurements beyond x/D = 60   is
YES.   Both U_0 and $\overline{u_{\mathfrak{C}}^2}^{\frac{1}{2}}$ vary as $x^{-2/3}$ and $y_{\frac{1}{2}}$ as
$x^{1/3}$;   and the profiles of both mean and fluctuating
velocities are similar.   The spreading parameter
$S = U_E/U_0 (dy_{\frac{1}{2}}/dx)$ is constant even for x/D < 60.

    Influence of body shape.   The findings presented
so far were expected, but it comes as a surprise that
S   should depend strongly on the shape of the wake-
generating body.   Table 9 reveals that   S   varies

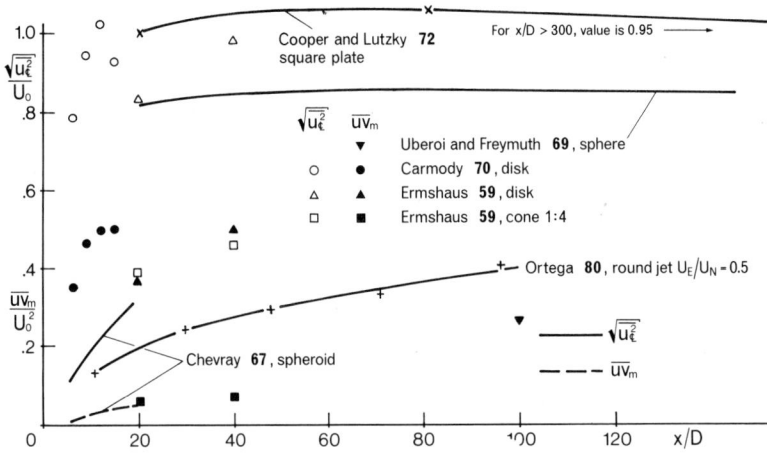

Fig. 28   Relative centre-line intensity and maximum
shear-stress in round wakes.

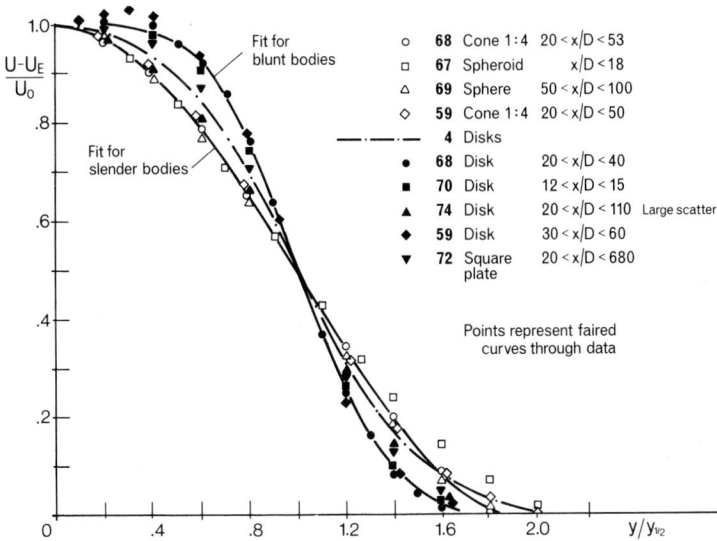

Fig. 29   Velocity profiles in self-similar round
wakes.

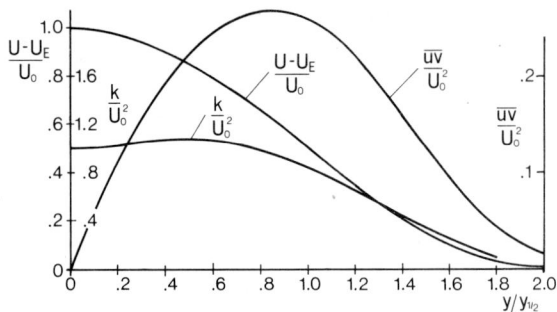

Fig. 30   Profiles of U, k, and $\overline{uv}$ in the self-
                 similar round wake behind a sphere: data
                 of Uberoi and Freymuth [69].

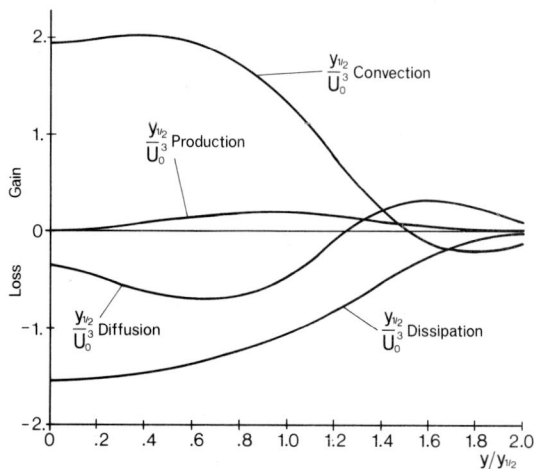

Fig. 31   Energy balance for the self-similar round
                 wake behind a sphere: modified data of
                 Uberoi and Freymuth [69].

from $\simeq .1$ for thin bodies to $\simeq .9$ for very blunt ones.
Along with this goes a drastic increase in the turbulence level relative to the defect velocity (see Fig.
28).   It is interesting to note that, in contrast to

plane wakes, the asymptotic value of $\overline{u_\mathfrak{C}^2}$ is approached
from below. Figure 29 shows that even the velocity
profile depends on the body shape. All the profiles
presented are taken from the self-preserving region
of the flows.

Turbulence field. Figure 30 shows Uberoi and
Freymuth's [69] similarity profiles for k and $\overline{uv}$,
and Fig. 31 their energy balance. The data were
found to be consistent. The author constructed a
shear-stress balance from Uberoi and Freymuth's
measurements which is shown in Fig. 32. There is
some evidence that the turbulence field in the far
wake depends much less on the body shape than does the
mean field. Figure 33 gives a plot of $\sqrt{u_\mathfrak{C}^2}/U_E$ versus x/D. The agreement of most of the data is
reasonable. Table 9 also lists the spreading parameter of the turbulence field,

$$S_t = \frac{U_E}{\sqrt{\overline{u_\mathfrak{C}^2}}} \frac{dy_{\frac{1}{2}t}}{dx} ,$$

where $y_{\frac{1}{2}t}$ is the radius at which $\sqrt{\overline{u^2}}$ is half of
$\sqrt{\overline{u_\mathfrak{C}^2}}$. Unfortunately, not enough data were available
to determine S_t for slender bodies, but the S_t's
for sphere and disk differ much less than do the corresponding mean-field parameters S. It is interesting to note that the value of $S_t = 1.21$ for disks is
close to the value of 1.26 for Townsend's [56] shear-
free layer discussed in Sec. 5.5. All these findings
indicate that the turbulence field is practically de-
coupled from the mean field and behaves therefore as
in shear-free layers. Examination of the energy balance shown in Fig. 31 offers an explanation for this:
the production, by means of which the mean field exerts

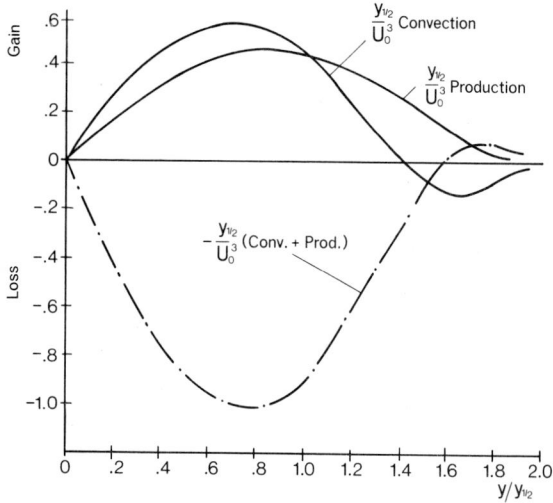

Fig. 32   Shear-stress balance for the self-similar
               round wake behind a sphere. Constructed
               from the data of Uberoi and Freymuth [69].

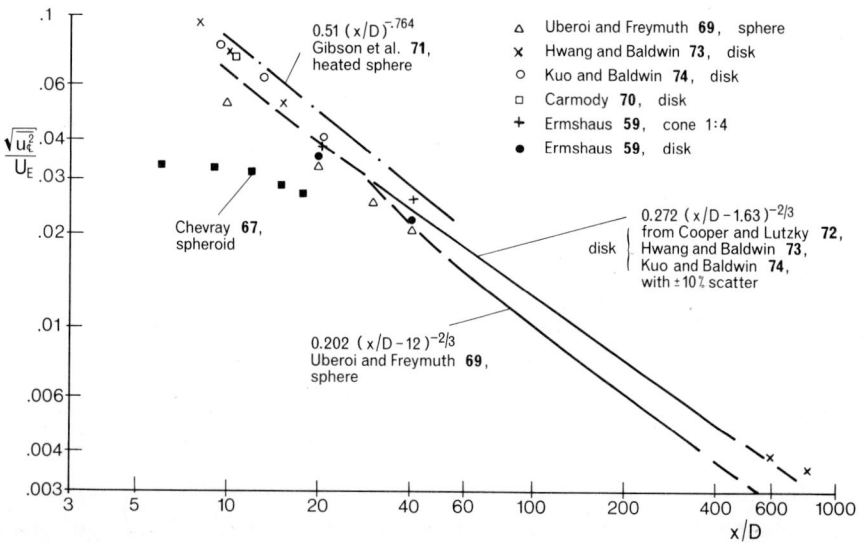

Fig. 33   Axial decay of $\sqrt{\overline{u^2}}$ in round wakes.

its influence, is much smaller than the other terms;
in shear-free layers it is zero.

Figures 31 and 32 show, respectively, that, in
contrast to jets for example (see Figs 14, 15, 18, 19),
the convection of both energy and shear stress is very
important in round wakes. This appears to be the
reason for the large "upstream effects" in these flows.

Reynolds number influence. A further common
feature of round wakes and shear-free layers is the
linear decay with x of γ_ℓ (Baldwin and Sandborn
[75], Mobbs [76]). This behaviour may be attributed
to the decreasing Reynolds number; it indicates that
the observed self-similarity can only be an approximate one. This makes it easier to accept the existence of different states of similarity behind different bodies. The decreasing Reynolds number may therefore be the basic reason for this apparent oddity. It
would be intriguing to conduct a round-wake experiment
with the free-stream velocity U_E varying as x^m,
where $m < -1/3$ (but $m > -2/3$). In an approximately
self-similar flow of this kind, the Reynolds number
rises with x (according to Table 2), and the far
wake should be independent of the body shape if the
above-stated hypothesis were true.

5.5 *Flow Behind Self-propelled Bodies (Shear-free Layers)*

The flow behind plane and axisymmetric self-
propelled bodies represents respectively a line and a
point source of turbulence in a uniform stream (see
Fig. 2). We are interested here only in the far region where the mean velocity is uniform and the shear
stress is zero. The plane case has been investigated
by Townsend [56] and Mobbs [76], the axisymmetric case

by Naudascher [77].

   Nature of the flow.   Because this flow may not
be so well known to the reader, a short account is
given first of Mobbs' findings on the nature of the
flow:

>   "A sharp, irregular boundary separates the tur-
>   bulent from the non-turbulent part of the flow.
>   Within the turbulent part, the intensity is
>   laterally homogeneous, so that the dependence
>   on y of the overall intensity is caused solely
>   by the intermittency.   The profiles of k/k_o
>   and the intermittency factor γ/γ_ℓ are there-
>   fore identical. γ_ℓ decreases linearly with
>   x.   The apparent outward spreading of the tur-
>   bulence is due entirely to the growing ampli-
>   tude of the turbulent bulges.   The volume of
>   the turbulent fluid decreases with increasing
>   x."

The flow seems therefore to be unique in that its
overall behaviour is that of spreading inhomogeneous
turbulence, while the turbulent part contracts and is
homogeneous.

   Self-similarity.   From Sec.3.2 we expect the
flow to be self-similar, i.e. the spreading parameter

$$S \;=\; \frac{U_E}{\sqrt{k_o}}\,\frac{dy_{\frac12}}{dx}$$

should be independent of x, and k/k_o and $\varepsilon y_{\frac12}/k_o^{\frac32}$
should be functions of $y/y_{\frac12}$ only (for definitions
see Fig. 2).   All three investigators found the k-
profile to be indeed self-similar; but there is little
consensus about the parameter S.   For the plane case,
Townsend's results give S = .75, but Mobbs' results

indicate a significant variation of S in the region
investigated.   For the axisymmetric case, the writer
has evaluated Naudascher's data and found that S is
approximately constant over a very limited range only
(15 < x/D < 25), where its value is about 0.4.

   Reynolds number effects.   In Sec. 3.2 we have
seen that the Reynolds number $\sqrt{k_o}\, y_{\frac{1}{2}}/\nu$ decreases with
x when k_o decays as x^{-n} and n is larger than 1.
In this case, the similarity region is restricted if
existent at all.   Townsend reports n $\simeq$ 1.2, but both
Mobbs and Naudascher report n $\simeq$ 1.7.   Because of the
homogeneity of the turbulent part of the flow, we
would expect the axial intensity decay to be similar
to that behind grids when the Reynolds number is still
high.   This notion is supported by Townsend's value
of n $\simeq$ 1.2.*   The high value of 1.7 reported by the
other two authors indicates that their flows may have
been subject to low Reynolds number effects.   This
is consistent with the lack of constancy of the parameter S and thus of true self-similarity.   Tennekes
and Lumley [79] offer another possible explanation for
the apparent lack of self-similarity: when the experimental excess (or deficit) momentum flux is not exactly
zero, the self-propelled component of the flow is soon
overshadowed by the momentum excess (or deficit) component.   It is indeed quite difficult to arrange that
the momentum flux is precisely zero in an experiment.

   k-*profile and* k-*balance.*   The profile of k/k_o
is shown in Fig. 34.   The results of Townsend and
Naudascher are virtually identical.   The energy bal-

---

* Comte-Bellot and Corrsin [78] give a comprehensive
review on grid-turbulence data.   They found that the
majority of the measurements supports a value of n$\simeq$1.2.

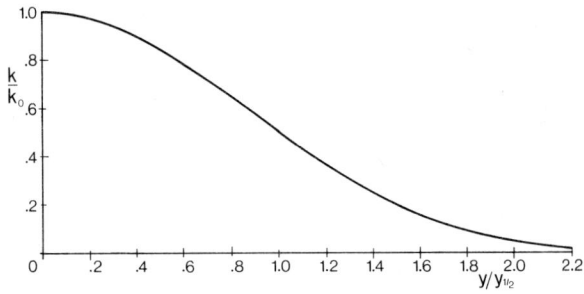

Fig. 34   k-profile for shear-free layers.   Data
               of Townsend [56] (plane case) and
               Naudascher [77] (axisymmetric case).

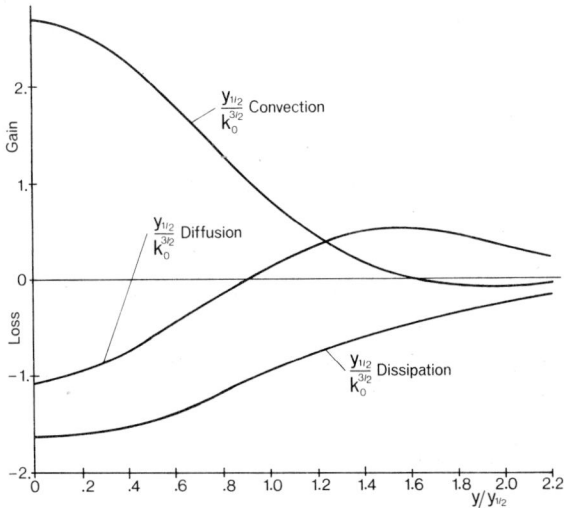

Fig. 35   Energy balance for the plane shear-free
               layer.   Data of Townsend [56], modified.

ance as constructed from Townsend's measurements is
shown in Fig. 35.   Townsend's dissipation values were
multiplied by a factor of 1.5 to make the diffusion
integrate to zero.   The diffusion was determined as
the difference of convection and dissipation.   In

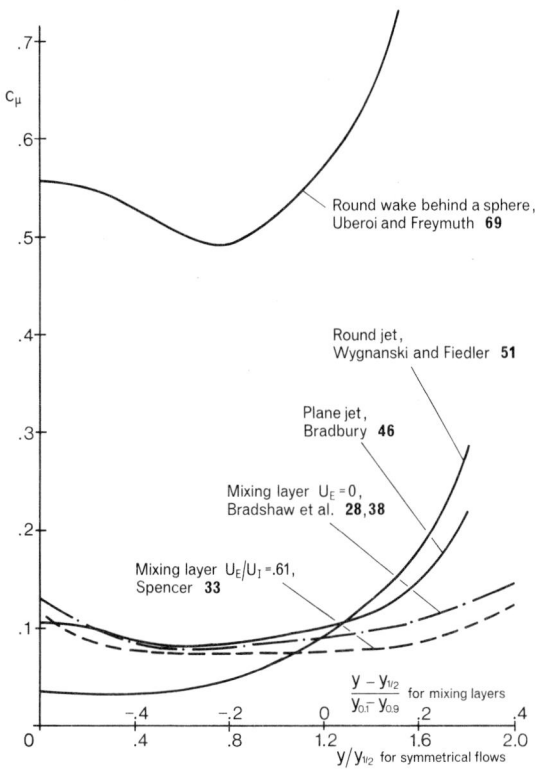

Fig. 36 c_μ across self-similar free turbulent boundary layers.

this shear-free flow the shear-stress production is, of course, zero.

5.6 c_μ and $\overline{uv}/k$ in Self-similar Free Boundary Layers

Figures 36 and 37 show respectively the cross-wise distribution of the parameters c_μ and $\overline{uv}/k$ appearing in the shear-stress hypotheses (2.5) and (2.6). The parameters were determined from the measurements of U, $\overline{uv}$, k and ε discussed above. Flows are included only in Figs. 36 and 37 for which

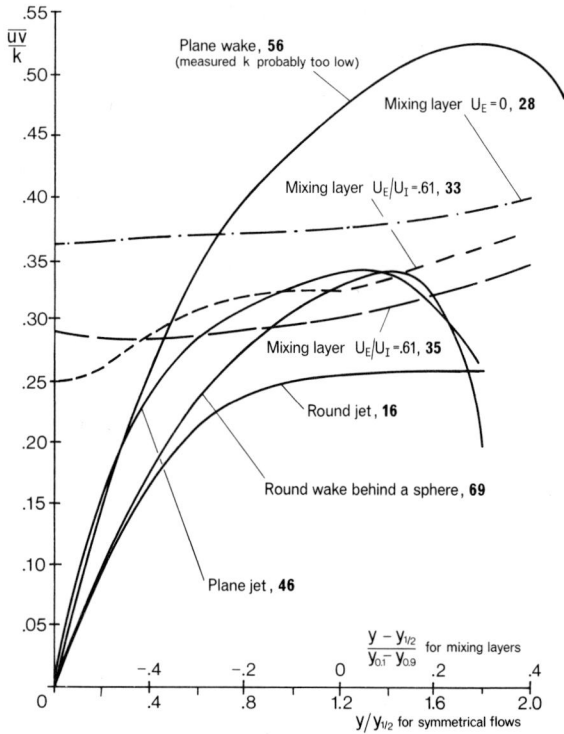

Fig. 37 $\overline{uv}/k$ across self-similar free turbulent
              boundary layers.

all the data were found to be consistent.   Figure 36
suggests that $c_\mu \simeq .09$ is a suitable average value
for mixing layers and for the plane jet (it is suitable
also for wall boundary layers, see Ng and Spalding
[15]) but that the round jet, and particularly the
round wake require different values.   From a more detailed study, Rodi [16] concluded that $c_\mu \simeq .09$ is
appropriate for nearly all boundary layers (both near
to and remote from walls) having strong strain.   The
round jet is an exception.

    Figure 37 shows that, when we exclude the somewhat suspect data of Townsend [56], $\overline{uv}/k \simeq .3$ is a suitable average value for flow regions not too close to symmetry lines or planes. This is the value adopted by Bradshaw *et al.* [13] for wall boundary layers. Figure 37 shows however that the scatter is approximately ±20% about this value. At a centre-line or plane, $\overline{uv}/k$ must, of course, go to zero. In this region, $\overline{uv}/k \simeq .3$ cannot therefore be a reasonable approximation.

6. REVIEW OF DATA FOR JETS ISSUING INTO A UNIFORMLY MOVING STREAM

Jet forgetfulness. Jets issuing into a uniformly moving stream cannot be self-similar because U_E/U_O varies with x; here therefore we are particularly interested in the streamwise flow development. The principle of "jet forgetfulness" (Sec. 3.3) suggests that, some way downstream of the nozzle, this development depends on the variable $(x-x_O)/\theta$ only and not on the velocity ratio U_E/U_N.* Both plane and round jet experiments confirm this. An example is given in Fig. 40 where Reichardt's [66] and Ortega's [80] actual data points of $y_{\frac{1}{2}}/\theta$ and U_E/U_O are plotted versus x/θ (x_O was assumed to be zero).

Variation of U_O *and* $y_{\frac{1}{2}}$. With the aid of the integral momentum equation (3.1), the non-dimensional width, $y_{\frac{1}{2}}/\theta$, can be expressed as

---

* The reader is reminded that the location of the virtual origin, x_O, does depend on the flow conditions at the nozzle.

W. *Rodi*

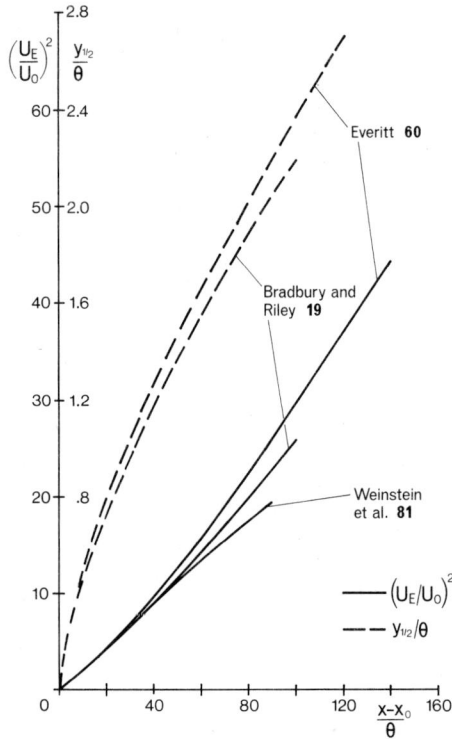

Fig. 38   Plane jet issuing into a uniformly moving
            stream.

$$\frac{y_{\frac{1}{2}}}{\theta} = \frac{1}{\left[(2\pi)^j \left(\frac{U_O}{U_E} I_1 + \left(\frac{U_O}{U_E}\right)^2 I_2 \right)\right]^{\frac{1}{j+1}}} \quad , \qquad (6.1)$$

where $I_1 = \int_O^\infty f\eta^j d\eta$ and $I_2 = \int_O^\infty f^2\eta^j d\eta$.

Equation (6.1) was used in this review to check the
consistency of the measurements.

For the *plane jet*, measured values of $y_{\frac{1}{2}}/\theta$ and

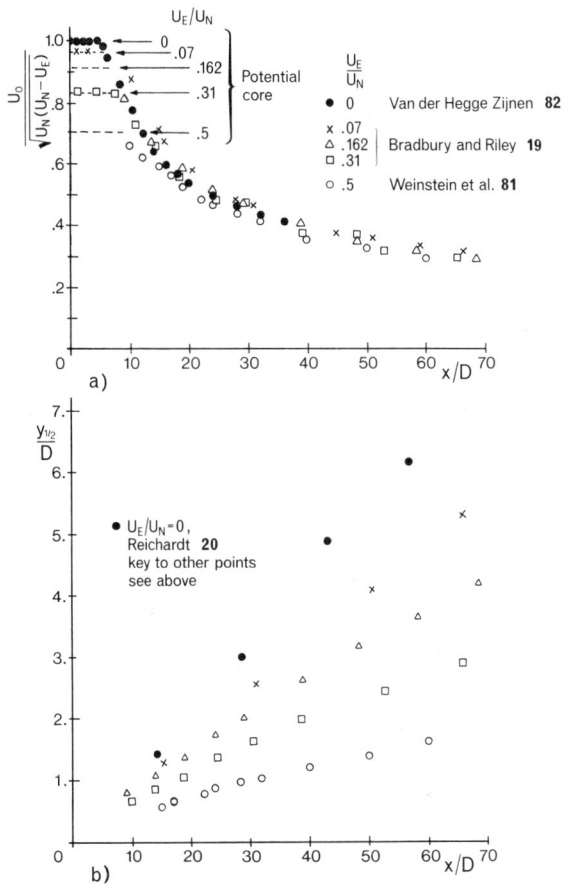

Fig. 39   Plane jet: near field.   a) velocity
           decay.   b) jet growth.

$(U_E/U_O)^2$ are plotted versus $(x-x_O)/\theta$ in Fig. 38.
Each experimenter's data satisfy equation (6.1). The
agreement between different experimenters is good
only at small values of $(x-x_O)/\theta$. At larger values,
Everitt's [60] data are probably closest to the truth
because he conducted the most extensive investigation.
At small values of U_E/U_N, the quantity θ/D is large

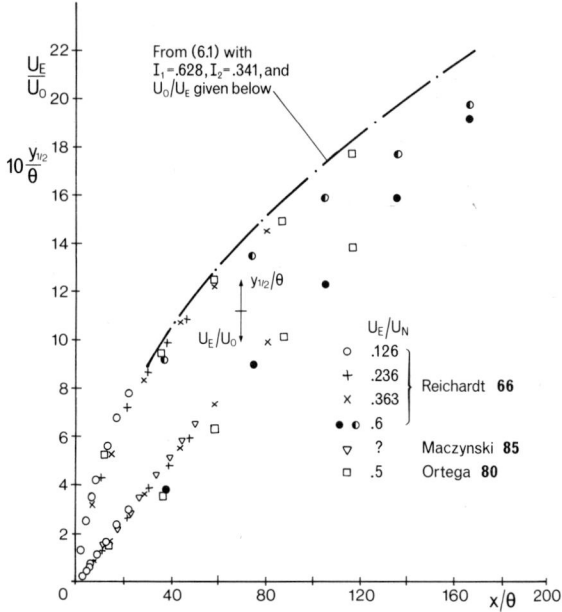

Fig. 40   Round jet issuing into a uniformly moving
               stream.

and Fig. 38 does not illustrate clearly the flow development in the near field.   Results for the near
field are therefore given in Fig. 39 where
$U_O/\sqrt{U_N(U_N-U_E)}$ and $y_{\frac{1}{2}}/D$ are plotted versus x/D.
With the above assumption of jet forgetfulness, and
of a uniform velocity profile at the nozzle exit,*
the velocity data should again fall on one curve (except in the potential-core region).   Figure 39 shows
that they do so fairly well.

---

* For uniform flow at the nozzle, the following relationship holds:

$$\frac{\theta}{D} = (\frac{\pi}{4})^j \frac{U_N}{U_E} \left(\frac{U_N}{U_E} - 1 \right)^{\frac{1}{j+1}} .$$

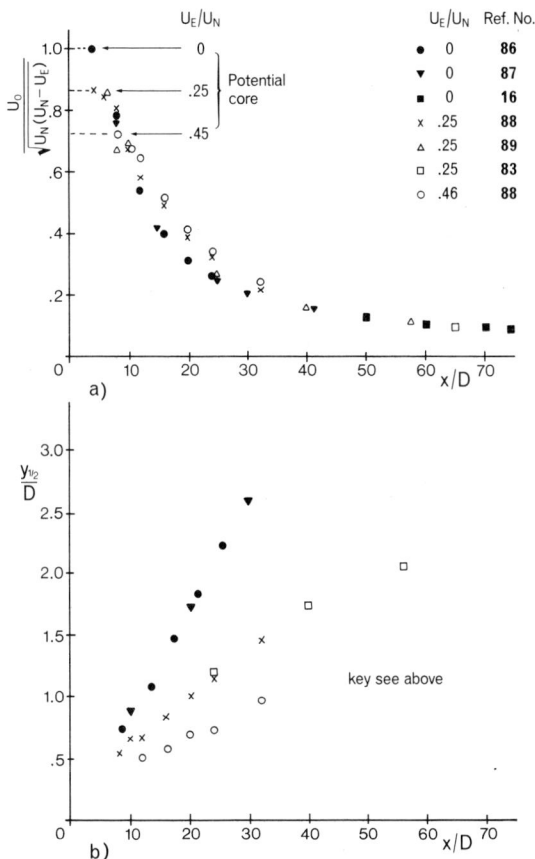

Fig. 41   Round jet: near field.   a) velocity
decay.   b) jet growth.

The corresponding results for the *round jet* are
plotted in Figs 40 and 41.   The consensus of different experimenters is fairly good.   At high values of
x/θ (Fig. 40), the measured values of $y_{\frac{1}{2}}/\theta$ do not
agree with $y_{\frac{1}{2}}/\theta$ as calculated from equation (6.1).
The measurements of U_o are probably more accurate
than those of $y_{\frac{1}{2}}$,   but they are somewhat suspect too.
The results of Forstall and Shapiro [83] and Curtet

and Ricou [84] are not included in Fig. 40 because,
for the higher values of U_E/U_N, their flows were
influenced noticeably by the duct which guided the
outer stream.

In both plane and round jets, the shape of the
velocity profile is approximately the same and does
not vary significantly with either U_E/U_N or x.

The asymptotic behaviour. The ratio U_O/U_E
falls steadily with x; and, at large distances,
where U_O/U_E is very small, the flow is expected to
assume wake-like character. For $U_O/U_E \ll 1$, the
similarity analysis predicts that the flow width should
vary as $x^{1/(j+2)}$ and the excess velocity U_O as
$x^{-(j+1)/(j+2)}$. The parameter

$$S = \frac{U_E}{U_O} \frac{dy_{\frac{1}{2}}}{dx}$$

should thus be a constant. The question is: did the
measurements extend far enough downstream to satisfy
the condition that $U_O/U_E < .06$ (see Sec. 5.4)? The
lowest values of U_O/U_E included here are 0.158 and
0.053 for the plane and round jet respectively. They
are probably not low enough for the establishment of
truly self-similar wake-type flow. The measurements
in wakes extended to much lower values (see Tables 8
and 9).

Nevertheless, Everitt's data testify that the
plane jet approaches the expected asymptotic state.
At his furthest station, S still continues to rise,
but its value of 0.087 is already close to the value
of 0.098 in wakes (Table 8).

The round jet, on the other hand, exhibits no
tendency to approach the expected asymptotic state.

For $x/\theta > 50$, U_O/U_E varies approximately as x^{-1}, and $y_{\frac{1}{2}}$ as $x^{\frac{1}{2}}$. The expected behaviour is $U_O/U_E \propto x^{-2/3}$ and $y_{\frac{1}{2}} \propto x^{1/3}$. This surprising result may possibly be explained by the fact that the Reynolds number decreases with x in this flow. The assumption of high Reynolds number, which is inherent in the similarity analysis, may thus be invalid. This explanation is supported by the finding that the variation of U_O and $y_{\frac{1}{2}}$ is identical to that expected in a laminar self-similar axisymmetric wake (Schlichting [90]).

Turbulence measurements. Turbulence quantities were measured by Everitt in the plane jet and by Curtet and Ricou and by Ortega in the round jet. In all the cases, $\overline{u_{\xi}^2}/U_O^2$ rose monotonically with x (see Ortega's results, Fig. 28). Everitt's energy balance shows that the average ratio of production to dissipation of turbulent kinetic energy decreases markedly as U_O/U_E decreases. Unfortunately, Everitt's turbulence data were found to be not very consistent and are therefore not presented here.

7.   SUGGESTIONS FOR FUTURE EXPERIMENTS

For most flows considered in this paper, satisfactory measurements of both mean and turbulence quantities could be identified;   and the general behaviour of the flows was found to correspond to that expected from the theory of Sec. 3.   However, in spite of the great number of existing data, some basic phenomena still lack a proper explanation;   and some flows have been rather neglected.   The article therefore closes with a few suggestions for future experimental work.

The influence of the apparatus layout and the intensity and scale of free-stream turbulence on mixing layers should be studied systematically, with U_E/U_I varying over its whole range (zero to unity). Such experiments would allow a definite decision on whether or not the velocity ratio U_E/U_I influences the turbulence structure. Another attempt should be made to establish a truly self-similar radial jet. A systematic study should be carried out on the effect of the body shape on plane and round wakes. Also, the turbulence quantities in the self-similar regions of these flows should be re-examined. Finally, the possible effect of the Reynolds number on shear-free layers, round wakes, and round jets in a moving stream should be investigated.

In laboratory flows, some of the quantities listed in Table 1 could not be measured with satisfactory accuracy (e.g. dissipation ε); and some could not be measured at all (e.g. terms involving pressure fluctuations). Data on these quantities would aid greatly the development of turbulence models solving transport equations for the Reynolds stresses. Methods should therefore be developed which allow these quantities to be measured reliably in laboratory flows.

ACKNOWLEDGEMENTS

The author wishes to thank Dr. R.J. Keller for helpful comments on this paper.

REFERENCES

1. Launder, B.E. and Spalding, D.B. "Mathematical
    models of turbulence", Academic Press, London
    and New York, 1972.

2. Launder, B.E., Morse, A., Rodi, W. and Spalding,
    D.B. "Prediction of free shear flows - A com-
    parison of the performance of six turbulence
    models", *Proceedings of the NASA-Langley Con-
    ference on Free Turbulent Shear Flows*, NASA-
    SP-312, 1973.

3. Abramovich, G.N. "The theory of turbulent jets",
    The M.I.T. Press, 1963.

4. Halleen, R.M. "Literature review on subsonic
    free turbulent shear flow", Stanford Univer-
    sity, Mech. Eng. Report MD-11, 1964.

5. Spalding, D.B. "Monograph on turbulent boundary
    layers", Imperial College, Dept. of Mech. Eng.
    Report TWF/TN/24, 1967.

6. Newman, B.G. "Turbulent jets and wakes in a
    pressure gradient", *In:* Fluid Mechanics of
    Internal Flow (G. Sovran, ed), Elsevier,
    p. 170, 1967.

7. Harsha, P.T. "Free turbulent mixing: a critical
    evaluation of theory and experiment", Arnold
    Engineering Development Center, Tennessee,
    Report AEDC-TR-71-36, 1971.

8. Hanjalić, K. and Launder, B.E. "A Reynolds stress
    model of turbulence and its application to thin
    shear flows", *J. Fluid Mech.*, $\underline{52}$, 609-638,
    1972.

9. Rodi, W. "Basic equations for turbulent flow in
    cartesian and cylindrical coordinates", Imperial
    College, Mech. Eng. Dept. Report BL/TN/A/36,
    1970.

10. Rodi, W. and Spalding, D.B. "A two-parameter
    model of turbulence, and its application to
    free jets", *Wärme- und Stoffübertragung*, $\underline{3}$,
    No.2, 85-95, 1970.

11. Kolmogorov, A.N. "Equations of turbulent motion
    in an incompressible fluid", *Izv. Akad. Nauk.
    SSSR, Seria fizicheska Vi*, $\underline{1-2}$, 56-58,
    1942 (English translation: Imperial College,
    Mech. Eng. Dept. Report ON/6, 1968).

12. Prandtl, L. "Über ein neues Formel-System für
     die ausgebildete Turbulenz", Nachr. Akad.
     Wiss., Göttingen, Math.-phys. KL. 1945, p. 6,
     1945 (with an appendix by K. Wieghardt).

13. Bradshaw, P., Ferriss, D.H. and Atwell, N.P.
     "Calculation of boundary-layer development
     using the turbulent energy equation", *J. Fluid
     Mech.*, 28, pp. 593-616, 1967.

14. Glushko, G.S. "Turbulent boundary layer on a flat
     plate in an incompressible fluid", *Izv. Akad.
     Nauk. SSSR, Mekh.* 4, 13, 1965.

15. Ng, K.H. and Spalding, D.B. "Turbulence model for
     boundary layers near walls", *The Physics of
     Fluids*, 15, No. 1, 20-30, 1972.

16. Rodi, W. "The prediction of free turbulent bound-
     ary layers by use of a two-equation model of
     turbulence", Ph.D. Thesis, University of London,
     1972.

17. Gartshore, I.S. and Newman, B.G. "Small-
     perturbation jets and wakes which are approxi-
     mately self-preserving in a pressure gradient",
     C.A.S.I. Transaction, 2, 101, 1969.

18. Spalding, D.B. "Spread of confined turbulent pre-
     mixed flames", *Seventh Symposium (International
     on Combustion*, Butterworth, London, 1958.

19. Bradbury, L.J.S. and Riley, J. "The spread of a
     turbulent plane jet issuing into a parallel
     moving airstream", *J. Fluid Mech.*, 27, 381-394,
     1967.

20. Reichardt, H. "Gesetzmässigkeiten der freien
     Turbulenz", *VDI-Forschungsheft*, 414, 1942.

21. Liepmann, H.P. and Laufer, J. "Investigations of
     free turbulent mixing", NACA TN. 1257, 1947.

22. Wygnanski, I. and Fiedler, H.E. "The two-
     dimensional mixing region", *J. Fluid Mech.*, 41,
     327-363, 1970.

23. Patel, R.P. "A study of two-dimensional symmetric
     and asymmetric turbulent shear flows", Ph.D.
     Thesis, McGill University, 1970.

24. Albertson, K.L., Dai, Y.B., Jensen, R.A. and Rouse
     H. "Diffusion of submerged jets", *Proc. Am.
     Soc. Civil Engrs.*, 74, 1751, 1948.

25. Mills, R.D. "Numerical and experimental investigations of the shear layer between two parallel streams", *J. Fluid Mech.*, <u>33</u>, 591-616, 1968.

26. Sunyach, M. and Mathieu, J. "Zone de mélange d'un jet plan; fluctuations induite dans le cône a potential - intermittance", *Int. J. Heat Mass Transfer*, <u>12</u>, 1679-1697, 1969.

27. Maydew, R.C. and Reed, J.F. "Turbulent mixing of axisymmetric compressible jets (in the half-jet region) with quiescent air", SANDIA Corp. Aerothermodynamics, SC-4764, 1963.

28. Bradshaw, P., Ferriss, D.H. and Johnson, R.F. "Turbulence in the noise-producing region of a circular jet", *J. Fluid Mech.* <u>19</u>, 591-624, 1964.

29. Sami, S., Carmody, T. and Rouse, H. "Jet diffusion in the region of flow establishment", *J. Fluid Mech.*, <u>27</u>, 231-252, 1967.

30. Miles, J.B. and Shih, J.S. "Similarity parameter for two-stream turbulent jet-mixing region", *AIAA J.*, <u>6</u>, 1429-1430, 1968.

31. Seban, R.A. and Back, L.H. "Velocity and temperature profiles in turbulent boundary layers with tangential injection", *Trans. ASME, J. Heat Trans.*, <u>84</u>, Series C, 45-54, 1962.

32. Sabin, C.M. "An analytical and experimental study of the plane, incompressible turbulent free-shear layer with arbitrary velocity ratio and pressure gradient", ASME paper No. 64-WA/FE 19, 1965.

33. Spencer, B.W. "Statistical investigation of turbulent velocity and pressure fields in a two-stream mixing layer", Ph.D. Thesis, University of Illinois, 1970.

34. Brown, G. and Roshko, A. "The effect of density difference on the turbulent mixing layer", Paper delivered at AGARD Meeting London, Sept. 1971.

35. Yule, A.J. "Two-dimensional self-preserving turbulent mixing layers at different free-stream velocity ratios", ARC 32 732, FM 4213, 1971.

36. Watt, W.E. "The velocity-temperature mixing layer", University of Toronto, Mech. Eng. Report TP-6705, 1967.

37. Gartshore, I.S. and Pui, N.K. "Turbulent mixing layers between parallel streams", Dept. of Mech. Eng., University of British Columbia, November 1971.

38. Bradshaw, P. and Ferriss, D.H. "The spectral energy balance in a turbulent mixing layer", NPL Aero Report 1144, 1965.

39. Bradshaw, P. "The effect of initial conditions on the development of a free shear layer", J. Fluid Mech., 26, 225-236, 1966.

40. Laurence, J.C. "Intensity, scale and spectra of turbulence in mixing region of free subsonic jet", NACA Report 1292, 1956.

41. Lassiter, L.W. "Turbulence in small air jets at exit velocities up to 705 feet per second," J. Appl. Mech., 24, 349-354, 1957.

42. Sami, S. "Balance of turbulent energy in the region of jet-flow establishment", J. Fluid Mech., 29, 81-92, 1967.

43. Castro, I.P. "A highly distorted turbulent free shear layer", Ph.D. Thesis, University of London, 1973 (available on microfiche from Dept. of Aeronautics, Imperial College, London SW7 2BX).

44. Flora, J.J. and Goldschmidt, V.W. "Virtual origins of a free plane turbulent jet", AIAA J., 7, 2344-2346, 1969.

45. Vagt, J.D. "Untersuchungen zur Turbulenzstruktur runder Freistrahlen", Interner Bericht am Herrmann-Föttinger-Institut für Strömungsmechanik, TU Berlin, 1970 (to be published).

46. Bradbury, L.J.S. "The structure of self-preserving turbulent plane jet", J. Fluid Mech., 23, 31-64, 1965.

47. Heskestad, G. "Hot-wire measurements in a plane turbulent jet", J. Appl. Mech., 32, 1, 1965.

48. Gutmark, E. "The two-dimensional turbulent jet", M.Sc. Thesis Technion-Israel Institute of Technology, 1970.

49. Robins, A. "The structure and development of a plane turbulent free jet", Ph.D. Thesis, University of London, 1971.

50. Gibson, M.M. "Spectra of turbulence in a round jet", *J. Fluid Mech.*, **15**, 161-173, 1963.

51. Wygnanski, I. and Fiedler, H.E. "Some measurements in the self-preserving jet", *J. Fluid Mech.*, **38**, 577-612, 1969.

52. Tuve, G.L. "Air velocity in ventilating jets", *Heating, Piping and Air Conditioning*, Jan. 1953.

53. Heskestad, G. "Hot-wire measurements in a radial turbulent jet", *J. of Appl. Mech.*, **33**, 417, 1966.

54. Fekete, G.I. "Two-dimensional self-preserving turbulent jets in streaming flow", McGill University Report M.E.R.L. 70-11, 1970.

55. Gartshore, I.S. "Two-dimensional turbulent wakes", *J. Fluid Mech.*, **30**, 547-560, 1967.

56. Townsend, A.A. "The structure of turbulent shear flow", Cambridge University Press, Cambridge, 1956.

57. Schlichting, H. "Über das ebene Windschattenproblem", *Ing.-Arch.*, **5**, 533, 1930.

58. Townsend, A.A. "The fully developed turbulent wake of a circular cylinder", *Australian J. Sci. Research.* **2A**, 451-468, 1949.

59. Ermshaus, R. "Eigentümlichkeiten turbulenter Nachlaufströmungen", Mitteilungen aus dem Max-Planck-Institut für Strömungsforschung No. 46, 1970.

60. Everitt, K. Ph.D. Thesis, University of London, 1971.

61. Alexopoulos, C.C. and Keffer, J.F. "Extended measurements of the two-dimensional turbulent wake", University of Toronto, Dept. of Mech. Eng. Report UTME-TP 6811, 1968.

62. Uberoi, M.S. and Freymuth, P. "Spectra of turbulence in wakes behind circular cylinders", *Physics of Fluids*, **12**, 7, 1359-1363, 1969.

63. Chevray, R. and Kovasznay, L.S.G. "Turbulence measurements in the wake of a thin flat plate", *AIAA J.*, **7**, 1641-1642, 1969.

64. Thomas, R.M. "Conditional sampling and other measurements in a plane turbulent wake", *J. Fluid Mech.*, **57**, 549-582, 1973.

65. Hinze, J.O. "Turbulence", McGraw-Hill, New York, 1959.

66. Reichardt, H. "Zur Problematik der turbulenten Strahlausbreitung in einer Grundströmung", Mitteilungen aus dem Max-Planck-Institut für Strömungsforschung No. 35, 1965.

67. Chevray, R. "The turbulent wake of a body of revolution", *Trans. ASME J. of Basic Engineering*, 90, Series D, 275, 1968.

68. Reichardt, H. and Ermshaus, R. "Impuls- und Wärmeübertragung in turbulenten Windschatten hinter Rotationskörpern", *Int. J. Heat Mass Transfer*, 5, 251-265, 1962

69. Uberoi, M.S. and Freymuth, P. "Turbulent energy balance and spectra of the axisymmetric wake", *Physics of Fluids*, 13, 9, 2205-2210, 1970.

70. Carmody, T. "Establishment of the wake behind a disk", *Trans. ASME J. of Basic Engineering*, 86, 869-882, 1964.

71. Gibson, C.H., Chen, C.C. and Lin, S.C. "Measurements of turbulent velocity and temperature fluctuation in the wake of a sphere", *AIAA J.*, 6, No. 4, 642-649, 1968.

72. Cooper, R.D. and Lutzky, M. "Exploratory investigation of turbulent wakes behind bluff bodies", David Taylor Model Basin Report 963, 1955.

73. Hwang, N.H.C. and Baldwin, L.V. "Decay of turbulence in axisymmetric wakes", *Trans. ASME, J. of Basic Engineering*, 88, Series D, 261-268, 1966.

74. Kuo, Y.H. and Baldwin, L.V. "Diffusion and decay of turbulent elliptic wakes", *AIAA J.*, 9, 1566, 1966.

75. Baldwin, L.V. and Sandborn, V.A. "Intermittency of far wake turbulence", *AIAA J.*, 6, 1163, 1968.

76. Mobbs, F.R. "Spreading and contraction at the boundaries of free turbulent flows", *J. Fluid Mech.*, 33, 227-239, 1968.

77. Naudascher, E. "Flow in the wake of self-propelled bodies and related sources of turbulence", *J. Fluid Mech.*, 22, 625-656, 1965.

78. Comte-Bellot, G. and Corrsin, S. "The use of a contraction to improve the anisotropy of grid-

generated turbulence", *J. Fluid Mech.*, **25**, 657-682, 1966.

79. Tennekes, H. and Lumley, J.L. "A first course in turbulence", The M.I.T. Press, 1972.

80. Ortega, J.J. "Characteristics of a turbulent round jet in a coaxial stream", M.S. Thesis, University of Iowa, 1968.

81. Weinstein, A.S., Osterle, J.F. and Forstall, W. "Momentum diffusion from a slot jet into a moving stream", *J. Appl. Mech.*, **78**, 437-443, 1956.

82. Van der Hegge Zijnen, B.G. "Measurements of the velocity distribution in a plane turbulent jet of air", *Appl. Sci. Res.*, Section A, **7**, 256-276, 1958.

83. Forstall, W., Jr. and Shapiro, A.H. "Momentum and mass transfer in coaxial gas jets", *J. of Appl. Mech.*, **17**, 399-408, 1950.

84. Curtet, R. and Ricou, F.P. "On the tendency of self-preservation in axisymmetric ducted jets", *Trans. ASME, J. Basic Engineering*, **86**, 765-775, 1964.

85. Maczynski, J.F.J. "A round jet in an ambient co-axial stream", *J. Fluid Mech.*, **13**, 597-608, 1962.

86. Corrsin, S. and Uberoi, M.S. "Further experiments on the flow and heat transfer in a heated turbulent air jet", NACA TN 1865, 1949.

87. Hinze, J.O. and Van der Hegge Zijnen, B.G. "Transfer of heat and matter in the turbulent mixing zone of an axially symmetric jet", *Appl. Sci. Res.*, Section A, **1**, 435-461, 1949.

88. Landis, F. and Shapiro, A.H. "The turbulent mixing of coaxial jets", *Proceedings of the Heat Transfer and Fluid Mechanics Institute*, 133-146, 1951.

89. Forstall, W., Jr. "Material and momentum transfer in coaxial gas streams", Ph.D. Thesis, Massachusetts Institute of Technology, 1949.

90. Schlichting, H. "Boundary layer theory", McGraw-Hill, New York, 1960.

NOMENCLATURE

Symbol	*Meaning*
a_1	empirical coefficient
c_μ	empirical coefficient
D	width/diameter of jet nozzle or wake-generating body
$\tilde{d}$	dimensionless diffusion of k $$\left(= v\,\frac{\overline{u_i u_i}}{2} + \frac{p}{\rho}\ \middle/\ U_o^3\right)$$
e	dimensionless kinetic energy ($= k/U_o^2$)
f	dimensionless longitudinal velocity, ($= (U-U_E)/U_o$)
g	dimensionless lateral velocity ($= V/U_o$)
h	dimensionless shear stress ($= \overline{uv}/U_o^2$)
j	exponent ($= 0$ for plane flow, $= 1$ for axisymmetric flow)
k	kinetic energy of turbulence ($= \frac{1}{2}\,\overline{u_i u_i}$)
L	length scale of turbulence
M	excess momentum flux
m	exponent of U_E ($U_E \propto x^m$)
n	exponent of k_o ($k_o \propto x^{-n}$)
p	fluctuating pressure
q	$(\overline{u^2} - \overline{v^2})/U_o^2$
Re	Reynolds number
S	spreading parameter
U, V, $(W=0)$	mean velocity components
U_o	velocity excess
u, v, w	fluctuating velocity components
x, y, z	coordinates
γ	intermittency factor

Symbol	*Meaning*
δ	characteristic flow width
ε	rate of dissipation
$\tilde{\varepsilon}$	dimensionless rate of dissipation $(= \varepsilon\delta/U_O^3)$
η	dimensionless distance $(= y/\delta)$
θ	momentum thickness
ν	kinematic viscosity
ρ	fluid density
σ	spreading parameter

Subscripts	*pertaining to*
E	external, outer stream
I	internal, inner stream
max, m	maximum
N	nozzle
t	turbulent
.1, $\frac{1}{2}$, .9	where velocity excess/deficit is .1, half, .9 of the maximum velocity excess/deficit
0, ℄	value at centre-line

DEVELOPMENTS IN LASER-DOPPLER ANEMOMETRY
AT IMPERIAL COLLEGE

by

J.H. Whitelaw

Department of Mechanical Engineering
Imperial College, London

ABSTRACT

A review of the principles and practice of laser-Doppler anemometry is provided as background information and to provide a basis for assessing the achievements of the present work on this topic. The achievements of the Imperial College group are identified under the headings of Contributions to understanding, Equipment and Applications. The required laser power, optical arrangements and the equivalence of their interpretation, the influence of particle size on signal intensity, correction factors for transit-time and gradient broadening, and the interpretation of the results obtained from different signal-processing systems are discussed under the first heading. Items of equipment referred to include integrated optical arrangements and frequency-shifting devices. Applications are discussed in terms of water flows, air flows and combusting flows: a small number of applications is discussed in greater detail in each category to bring out the relative advantages and remaining difficulties associated with the technique.

1. INTRODUCTION

The topic of the review is laser-Doppler anemometry and the contributions which have been made to its development and application in the Mechanical

Engineering Department of Imperial College.   The work
began early in 1969 and this review is concerned with
that carried out between then and December 1973.

The individual projects have been varied in charac
ter but there has been a common theme which exerted a
unifying and controlling influence on the work.   The
early stages of the work were concerned with instrument
development and led to the design and construction of
integrated optical units as well as to a basic understanding of the principles and practice of laser-Dopple
anemometry.   Several applications were investigated in
these early stages but these were of a demonstrative
nature, rather than serious fluid dynamic studies.
While such development and feasibility experiments have
continued into later stages of the work, progressively
greater effort has been devoted to the investigation of
fluid-dynamic phenomena in an effort to provide measurements of direct value to design and to the testing of
design procedures.   The unifying theme has been, therefore, to develop instrumentation capable of measuring,
with accuracy, velocity and related properties in a
range of practically relevant flow configurations and,
thereby, to aid designers.   This aim required that the
principles and practice of the instrumention be well
understood.

The present contributions can be viewed more readi
in the context of the contributions of other groups wor
ing in the same field over the same period.   The papers
which have stemmed from our efforts represent the third
generation of contributions relating to laser-Doppler
anemometry.   The first, dating from 1964, was concerned
with the realisation and simple demonstration that the
Doppler shift of laser light provided, in principle, a

convenient means for measuring velocity. The second
generation applied the technique to a wider range of
flow configurations but, in general, made use of non-
integrated optical systems and commercially available
frequency analysers; the measurements were, therefore,
largely limited to mean velocity and to indications
of turbulence intensity. The third generation of
contributions has involved the development of integrated optical systems, specially designed electronic-
signal processing arrangements and a substantial im-
provement in the understanding of the design features
necessary to allow measurements of turbulent-flow
properties with small error. Many demonstrative experiments and a few more extensive fluid-dynamic investigations have been reported. Also during this
third generation of contributions, several commercial
instruments have become available and have stemmed
largely from the efforts of research groups such as
those at the Atomic Energy Research Establishment
(Harwell), Arnold Research Organisation, General
Electric Corporation, Royal Radar Establishment and
Imperial College. It is to be expected that the
development of such 'off-the-shelf' instrumentation
will promote a still greater emphasis on fluid-dynamic
investigations in the near future.

The format of this article has been chosen to help
the reader identify the contributions of the group at
Imperial College and to view them in the context of
present knowledge. Most of the material derives from
published papers and from lecture notes prepared for
post-experience courses on laser-Doppler anemometry
presented at Imperial College over the past two years
(a more comprehensive version of those notes is now in

preparation in book form by F. Durst, A. Melling and
myself).

The sequence of topics in the present article is
as follows. Section 2 presents a brief discussion of
measurable flow properties and is followed by a longer
section concerned with the meaning of laser-Doppler
anemometry, the identification of the components which
make up a laser-Doppler anemometer and brief comments
on possible arrangements which may be employed; Sectio
3 is. intended to provide a foundation for the succeedir
three sections which review the more readily identifiable contributions of the Imperial College group. The
topics of these three sections are, respectively, contributions to the understanding of the principles and
practice of laser-Doppler anemometry, to equipment design and to the application of the technique to velocit
measurement in a range of flow configurations.

2.   MEASURABLE FLOW PROPERTIES

In principle, the laser-Doppler anemometer can
measure all of the properties indicated below:*

fundamental measurements: $\hat{U}_i$, $i = 1$ to 3

derived measurements: $\overline{U_i}$; $\overline{u_i^2}$, $\overline{u_i^3}$, $\overline{u_i^4}$ *etc.*,

$$P(U_i);   E(U_i);$$

$$\overline{u_i u_j},   \overline{u_i u_j^2},   \overline{u_i u_j^3} \text{ \textit{etc.}};$$

$$\overline{u_{i_A} u_{j_B}}   \overline{u_{i_A} u_{j_B}^2}   \overline{u_{i_A} u_{j_B}^3} \text{ \textit{etc.}}$$

As is indicated in Section 6, however, not all of these

---

* A definition of the symbols may be found on pp. 220-2

properties have been measured.   There are three main
reasons for this:

  (i)   the relatively short time that has elapsed
        since laser-Doppler anemometry was conceived
        as a technique for flow measurements.

 (ii)   fundamental limitations of laser-Doppler
        anemometry.

(iii)   the practical relevance of the flow proper-
        ties indicated above.

The following paragraphs expand upon the second and
third reasons.

As is well known, the principle of operation of
laser-Doppler anemometry necessitates the use of a
fluid which is not opaque and which contains particles
to scatter light.   Fluids such as blood, for example,
although containing particles (in the form of blood
cells) are not sufficiently transparent to allow
measurement if the light is required to pass through
more than approximately 300 μm of blood.   The need for
particles reminds us that it is the *particle* velocity
which is measured and that, in many cases, the particle
is required to follow the flow.   The need for particles
also means that, in most practical cases, the signal
will be intermittent.   This intermittency may be large
and then it is not possible to measure either the instantaneous velocity, $\hat{U}$, or the energy spectra $E(u_i)$.

Of the flow properties indicated above, the mean
velocity and turbulent stresses are generally considered to have direct practical relevance.   The other
properties assist understanding of turbulent flows and
the development of turbulence models.   It is unlikely
that correlations of order greater than four can, in
the near future, aid our understanding of turbulence or
that correlations of order greater than three will be

valuable for turbulence modelling.   Similarly, two-
point correlations are of limited applicability.   The
above considerations suggest that the emphasis should
be placed mainly on the measurement of $\overline{U}_i$ and $\overline{u_i u_j}$;
a considerably smaller number of researches can justify
measurements of other correlations.   Nevertheless, as
will be shown, measurements of higher order single-
point correlations can be obtained with comparatively
little effort using laser-Doppler anemometry.

3.   THE MEANING OF OPTICAL ANEMOMETRY

3.1   Preliminary Remarks

Figure 1 indicates the meaning of laser-Doppler
anemometry: in the left-hand diagram by reference to
sound waves and an aeroplane and, on the right, by
reference to a small particle and light waves.

In the case of the aeroplane, a sound wave transmitted from A will be received by a moving aeroplane

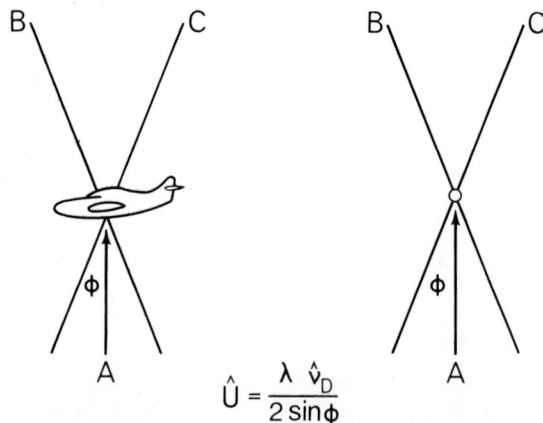

$$\hat{U} = \frac{\lambda \ \hat{v}_D}{2\sin\phi}$$

Fig. 1

at a different frequency.   If it is then re-trans-
mitted to B,   B   will receive a third frequency and
C   a fourth.   If the frequency received at C   is
subtracted from that at B,   the heterodyne signal is
directly proportional to the instantaneous velocity
of the aeroplane and the constant of proportionality
may be determined if the geometry of the system and
the original frequency (or wave length, λ) are known.

The same principle applies if the sound waves are
replaced by light waves and the aeroplane by a particle
The heterodyne signal will be directly related to the
instantaneous velocity of the particle by the formula
in Fig. 1.   The wave length refers to the wave length
of the light transmitted at A   and the angle refers
to the half angle between the light beams.

The linearity of the relationship between the instantaneous velocity and the instantaneous frequency
shift is particularly important.   It may be contrasted
with the non-linear relationship which exists with the
hot-wire anemometer.

More detailed explanations of the principles of
laser-Doppler anemometry have been given, for example,
by Durst and Whitelaw [1] and by Durst, Melling and
Whitelaw [2].   The brief explanation provided above in
terms of the Doppler effect may also be provided in
terms of wave theory or by reference to interference
fringes.

The constituent components of a laser-Doppler
anemometer are indicated on Fig. 2.   They consist of
the light source, an optical arrangement and a signal
processing system.   Of course, light scattering particles are also required in the measuring control volume.   Each of these components will be discussed in

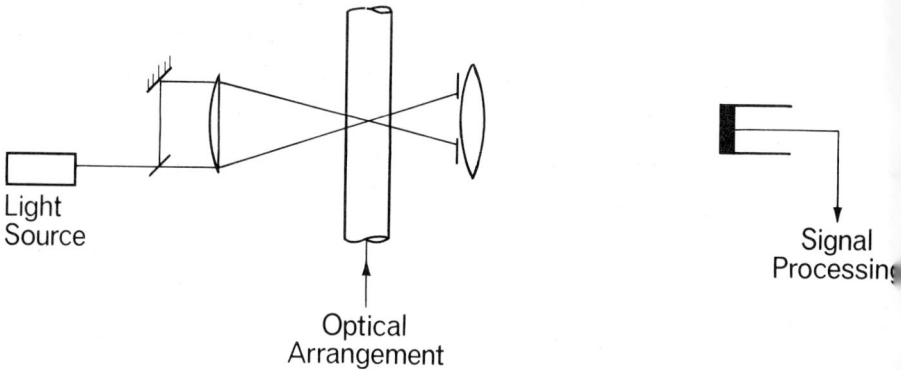

Light
Source

Optical
Arrangement

Signal
Processing

Fig. 2

the following sections.

It should be noted that the particular optical
configuration indicated in Fig. 2 corresponds to a system similar to that shown in Fig. 1 but with the signal
originating at   BC   and being received at   A.

3.2   Light Sources

In principle, any form of light source can be used
for Doppler-shift anemometry.   In practice, however,
laser light sources are exclusively used.   The reasons
stem mainly from the ability of a laser to provide a
sufficiently intense beam of small diameter.

Durst and Whitelaw [3] have shown that the required intensity of light is proportional to the fluid
velocity to be measured.   The actual intensity required will depend on the signal processing system employed.   An approximate figure of   0.50 mW   of He-Ne
laser power per m/s of velocity is useful as a guide.

The coherence of laser light is also important in
that it allows the measurement of a signal which will
include only one frequency unless it has been broadened

by turbulence or by measurement errors.   Similarly,
it is very useful to have a beam which diverges by a
small angle: the He-Ne laser beam diverges by only
1.7 milliradians.   The angle will, however, be effectively larger in situations such as large combustion
configurations where substantial refractive index
gradients are present.

3.3   *Optical Arrangements*

Figure 3 indicates three optical arrangements
which can be used to measure velocity.   The first
arrangement reminds us that we can measure velocity by
having two beams of light a known distance apart and
observing the passage of particles across the two light
beams.   Of course, this system is inconvenient in
that particles will not always cross the two light
beams.   It is also inconvenient in that the two light
beams are a finite distance apart and frequency response will be correspondingly low.   It can only be
used for measurements in low-turbulence flows.   The
cross-correlation method indicated by Fig. 3b will only
provide useful measurements when the two beams pass
through the same eddy;   the measurements cannot be local
since the two beams are a finite distance apart.

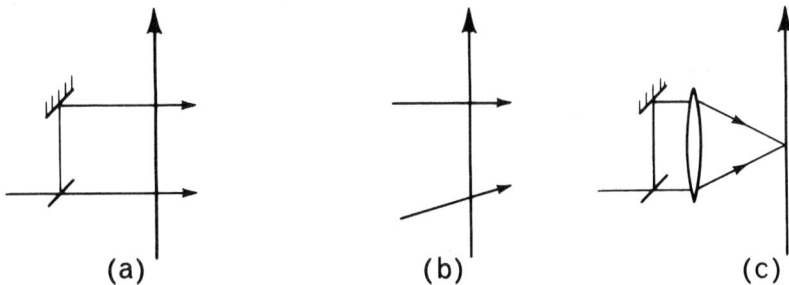

Fig. 3   Three optical methods for measuring velocity.

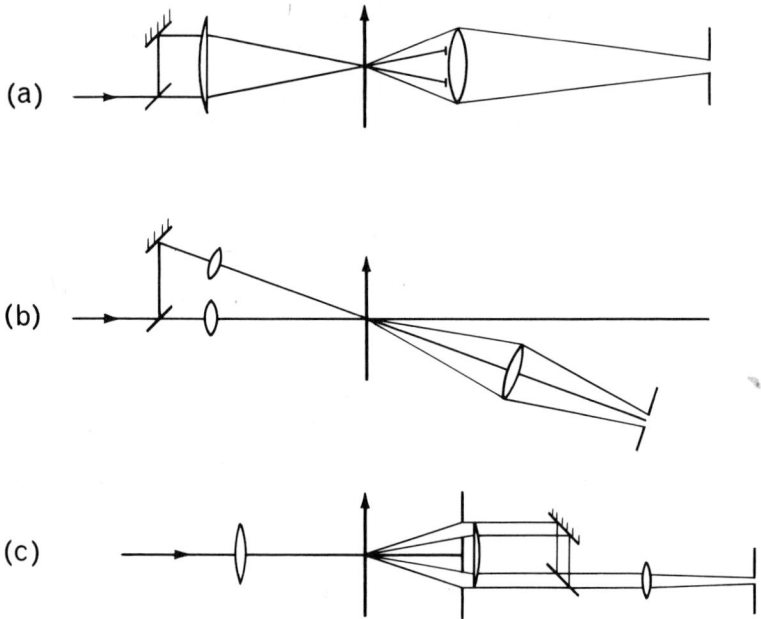

(a)

(b)

(c)

Fig. 4

In contrast to the above arrangements, Fig. 3c
indicates the interference of two light beams and this
is an essential feature of laser-Doppler anemometry.
It allows a local measurement, usually with a control
volume of the order of 1.0 mm in length and 0.1 mm
in diameter.

Figure 4 indicates three different geometrical
arrangements which can be used for laser-Doppler anemometry.   Each can be operated in a backward-scatter
form thereby allowing all the equipment to be arranged
on one side of the measuring control volume.

Figure 4a illustrates the so-called "fringe" system shown in Fig. 2.   This configuration is suited to
flow configurations where the particle concentration is
low and particles are not present in the measuring control volume at all times.   It is, therefore, suitable

for almost all practical measurements with laser-Doppler anemometry.   It should be noted that the siting of the light-collecting lens is particularly important.   The reader is referred to the paper by Durst and Whitelaw [3] for further detailed information.

The arrangement shown in Fig. 4b was used for most of the early work with laser-Doppler anemometry. It is characterised by a reference light beam which passes directly onto the photo-cathode.   In the example shown, the reference beam has also passed through the measuring control volume.   The light waves from this reference beam are heterodyned with the scattered light waves from the second light beam, the latter normally being nine or ten times more intense than the reference beam.   Drain [4] suggests that the reference mode provides a better signal-to-noise ratio (if operated coherently) in those flow configurations where there are many particles in the measuring control volume at any one time.   The minimum number of particles required in the control volume is not yet established but is probably greater than 10.   This arrangement offers advantages, therefore, only in a very limited range of flow configurations.

The single-beam mode of operation (Fig. 4c) corresponds to the example of Fig. 1 and may be thought of as the reverse of the fringe mode referred to in Fig. 3a.   Although no real fringes exist in the measuring control volume, virtual fringes do exist on the photo-cathode and the system is, therefore, very similar to the fringe mode.

3.4   *Signal Processing Systems*

Figure 5 indicates the nature of the signal ob-

J.H. Whitelaw

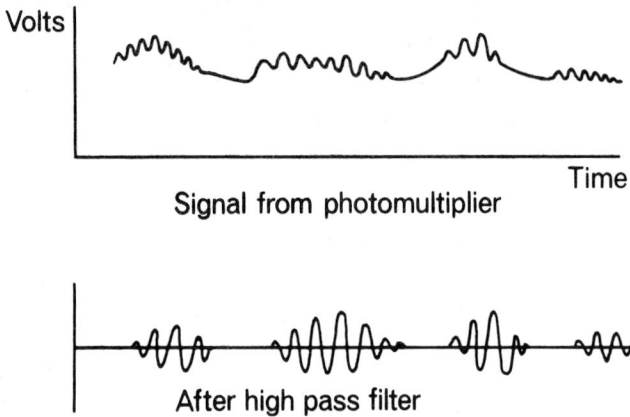

Signal from photomultiplier

After high pass filter

Fig. 5

tained from a photomultiplier before and after high-
pass filtering.   The particular signal shown is one
that would result from a very high particle concentration;   in many real flows, it may be expected that
the signal will be present for less than one per cent
of the time.   It is seen that the signal contains low
frequency and higher-frequency components.   These
are often referred to as dc and ac signals and,
although this terminology is imprecise, it is convenient and usual.   In fact, the ac signal refers to
the high frequency variations which correspond to
Doppler signals, *i.e*. those which we wish to measure.
The dc signal corresponds to the lower frequency component representing the passage of a particle through
one of the light beams; it contains no information
about the instantaneous velocity.   The frequency of
the signal within each envelope is proportional to the
instantaneous velocity component and the magnitude of
this signal depends on the particle size, its location
in the measuring control volume and the correct optical

alignment.

The signal may contain the effects of "shot" noise
and electronic noise and, in many real circumstances,
the quality of the signal itself may be low and the
signal-to-noise ratio at any instant may not be much
greater than unity. In these circumstances, it is
particularly important to ensure that all components
of the laser-Doppler anemometer are correctly matched.
In particular, it is important to note that, generally,
the poorer the signal-to-noise ratio the more expensive the signal processing system required. It is
therefore very important to ensure that the optical
arrangement is correctly designed.

In evolving the design, the laser power must be
calculated as indicated above. The ratio of particle
size to fringe spacing is also important and the optical arrangement should be designed to provide a fringe
spacing which is roughly equal to, or slightly greater
than, twice the mean particle diameter. The intensities of the two transmitted light beams should be
equal if the light is collected on the bisector of the
transmitted beam. Alignment is particularly important
as is a precise knowledge of the measuring control volume and its magnification upon the photocathode. The
light collecting system must, therefore, also be carefully designed.

It should be noted that the sign of the velocity
cannot be deduced from the signal unless means are provided for changing the frequency of one of the light
beams. Such an arrangement may involve the use of a
rotating grating, Denison and Stevenson [5], a Bragg
cell, Briard and Denham [6] or a Kerr cell, Drain and
Moss [7].

Three specific forms of signal processing are
examined in some detail in the succeeding paragraphs.
Topics such as the Fabry-Perot, the photon correlator
and filter banks are not discussed, however, so brief
mention is made of them here.   The Fabry-Perot is
known to operate well in flows with Mach numbers in
excess of 0.5 and its basic principle of operation
is similar to that of frequency analysis.   A number
of workers have discussed the feasibility of tracking
forms of Fabry-Perots but, so far, none have been produced or tested.   Photon correlators offer the potential advantage of operating with high velocity flows
and low power lasers: the direct output is, however,
the auto-correlation function and the precision of its
definition depends on the number of storage banks used.
This auto-correlation may be transformed, normally
through a fast-Fourier-transform computer, to a probability function which, in turn, may be transformed
into mean and rms information.   The precision of this
technique and the transforms has not yet been quanti-
fied.   The filter bank has the characteristics of
frequency analysis but, instead of just one filter, it
operates with a large number - typically about 50.
It, therefore, offers the possibility of making use of
50 times more signals than frequency analysis and this
is likely to be particularly important for measurements
in practical flow configurations such as industrial
furnaces.

The principle of *frequency analysis* requires the
use of a filter window, with effectively variable centre
frequency, which is swept across the signal at different
frequencies to produce a probability function.   The
band width of the filter is, of course, important and

$$\bar{v}_D = \frac{\int v_D P \, dv_D}{\int P \, dv_D}$$

$$\overline{(v_D - \bar{v}_D)^2} = \frac{\int (v_D - \bar{v}_D)^2 P \, dv_D}{\int P \, dv_D}$$

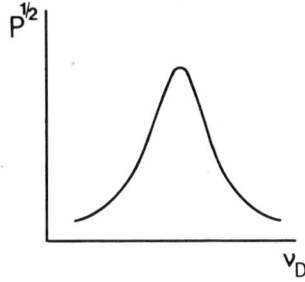

Fig. 6

those signals which do not occur at the filter setting
are rejected. The process does not, therefore, make
efficient use of available signals and may require a
significant time to build up a probability function.

The probability function thus obtained must, on
many occasions, be digitised and processed on the computer to obtain values of mean frequency, rms frequency
and higher-order correlations. Relevant equations are
shown on Fig. 6 and indicate that frequency analysers
can conveniently lead to values of mean velocity and
the one-point correlation functions, $\overline{u_i^2}$, $\overline{u_i^3}$, $\overline{u_i^4}$
etc.; these can readily be obtained from a probability
distribution. At Imperial College this technique has
been used in many flows where the signals are of poor
quality. In several combustion configurations, including a plasma jet, an industrial furnace and a blunt-
body-stabilised flame we have found that this technique
can yield measurements rather more easily than the available alternatives. It is, however, inefficient due to
the likely occurrence of many signal frequencies while
the filter sweeps the frequency range. Moreover, obtaining values of mean frequency and quantities corresponding to the normal stresses can be time consuming
and is often imprecise. The technique is particularly
imprecise when the mean frequency tends towards zero

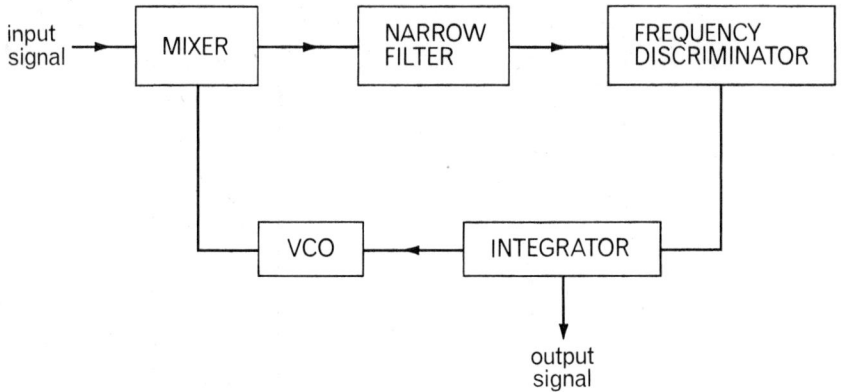

Fig. 7

while the range of frequencies is finite;   frequency
shifting of one of the light beams can, however,
readily remove this difficulty by providing a finite
mean frequency at zero velocity.

The principle of operation of the *frequency tracking demodulator* is shown in Fig. 7.   The input signal
is mixed with a voltage-controlled oscillator signal
and fed through a narrow-band filter to a frequency
discriminator.   This discriminator provides frequency-
to-voltage conversion and, after integration, an output
signal proportional to the input signal is produced as
a voltage.   This output voltage can be treated in a
manner identical to that of the hot-wire signal.

Although a frequency tracker provides what is ap-
parently a real-time signal, this signal can truly be
real time only if a particle is present in the measuring control volume at all times.   To deal with the non-
continuous nature of the signal, commercially available
frequency trackers provide "drop-out" control which
holds the last known frequency until a new signal appears.   This device does not, in principle, lead to
error if integration times are long enough to ensure

that the artificial signals are statistically random. Uncertainty in the length of time required to obtain this result renders frequency trackers less reliable in those flow configurations where drop out is greater than, say, 20%.

A practical limit of frequency trackers is the upper frequency limit of 50 MHz available in commercially available instruments. There is also a limit on the rate of change of frequency and this may be different for different instruments. Although systematic errors can arise from this source they are likely to be small.

Experience has shown that frequency trackers do not work well unless the particle concentration is high and does not vary significantly with time. This requirement effectively limits the use of frequency-tracking demodulators to laboratory conditions. In addition, frequency trackers suffer from electronic noise that usually reaches significant levels at the lower end of each frequency range.

Period timing offers many advantages in laser anemometry and is likely to become more important in the future. It operates on the simple principle of measuring the frequency of each signal and treating these statistically. The counter can be triggered at a finite slope and discrimination level, thereby initiating the counting of "zero crossings", *i.e.* times at which the trace of the signal crosses zero. It is possible to program counters to count a number of these crossings and to divide the total time taken by the number of crossings to produce a period and, therefore, a signal frequency proportional to instantaneous velocity. If this is done a large number of times, a

statistically significant value of mean velocity and
the corresponding rms component can be obtained. The
number of counts required depends on the property to
be evaluated and on the turbulence intensity.   On
many occasions, the required number exceeds 10,000 and,
if the particle concentration is small, this may require a long time.   The statistical nature of turbulence is, however, emphasised by this technique.

It is necessary to ensure that the count does not
pass from one signal envelope to another and that zero
crossings due to noise are not treated as useful signal
Logic circuitry is required to minimise possible errors
from this source.   In addition, the individual counts
must be averaged in a manner which ensures that errors
are not introduced, for example, by sampling more fast
moving particles than slow moving ones.

Period timing can provide values of mean velocity,
and correlations of the instantaneous velocity but not
in real time.   The statistical processing procedure
ensures that a value can only be provided at the end
of the required number of counts.   Variations in particle concentration do not influence this procedure and
it can also be used to reduce the sensitivity to flow
signals produced by very large or very small particles
or by particles moving through the edge of the control
volume.   This procedure of discrimination can help to
minimise the dimensions of the measuring control volume
but, in most cases, high values are required to ensure
that errors due to noise are minimised.

In principle, the technique is suited to highly
turbulent flows but the low frequency signal, referred
to previously, must be removed with care lest part of
the Doppler signal also be removed.   The noise-cancel-

ling procedure recommended by Hiller and Meier [8] and the use of frequency shifting, Baker, Hutchinson and Whitelaw [9,10] are likely to play increasingly important roles for this purpose.

3.5 *Particles*

Particles are of course necessary to provide a Doppler signal. They occur in nature but the mean diameter of particles normally present in the atmosphere is approximately two orders of magnitude lower than that needed for laser-Doppler anemometry. Particles in the atmosphere, suitable for laser-Doppler anemometry are present only in small numbers and their use alone would require a comparatively long time to obtain a measurement.

The criteria for the choice of seeding particles are generally that they should be of a size and density which allows them to follow the flow. Of course, there are situations where it is the *particle* velocity which is sought but in most cases, it is the flow velocity which is required. In the latter case, a useful guide for low subsonic velocities, is that a 1 μm diameter particle in air will allow a frequency response of around 10 kHz whereas a 10 μm diameter particle will allow a frequency response of only 700 Hz. Both figures quoted in the previous sentence assume a loss of precision in rms quantities of 1%.

Where seeding is provided, it is particularly important to ensure that the seed is non-toxic. At the same time, it should be realised that inhalation of small particles of any type may be bad for health. The reader is referred to the report by Melling [11] and to the paper by Melling and Whitelaw [12] for further information of this topic.

4. CONTRIBUTIONS TO THE UNDERSTANDING OF PRINCIPLES AND PRACTICE

It is particularly difficult to identify contributions which have been made to the understanding of the principles and practice of laser-Doppler anemometry. Unlike the more readily identifiable contributions of the succeeding two sections, it can be argued that the topics of this section have contributed more to our *own* understanding than to the understanding of others. Although in some cases this may be so, it is hoped that the selection of topics discussed here will prove helpful to many readers.   They are considered in the same order as the review of the previous section and represent amplifications and clarifications of points raised.

4.1 *Laser Power*

Lasers are available in discrete power ranges and, although it had been implicitly recognised that Argon lasers, with powers greater than 100 mW, should be used at higher velocities no guidance was available as to the power requirements and their variation with flow velocity, particle diameter *etc.*.   Durst and Whitelaw [13] have attempted to correct this omission.   They recommend the use of the following formula as a guide to the minimum laser power, P_ℓ, needed:

$$P_\ell \geq \frac{10^4}{4\pi^2} \frac{d_1^2 \ (h\overline{v}_o) \ \overline{v}_D}{\eta_q \ \eta_c \ Q_{scat} \ d_p^2 \ N_{ph}}$$

This formula suggests that the required laser power is

directly proportional to the Doppler frequency, the
control volume dimensions and the frequency of the
transmitted light: it is inversely proportional to
the collection efficiency of the light collecting system, the quantum efficiency of the photocathode, the
scattering coefficient obtained from Mie theory, particle radius squared and the number of fringes in the
measuring control volume. The analysis leading to the
above proposal is based on the timing of a single period
and assumes ideal band-pass filtering. The constants
in the equation contain the assumption that the number
of electrons required to define the signal over this
single period is 100.

In practice, the assumptions indicated in the last
two sentences of the above paragraph must be regarded
as idealisations. Available filters do not have ideal
characteristics and can never be preset to remove all
signals other than that under consideration. Similarly, the signal processing arrangement may require
more than 100 electrons per wave period and optical
components normally cause finite losses of light intensity. Thus the equation will give rise to a laser
power which may be significantly lower than that required in practice; it does, however, correctly represent the relationship between the required laser power
and the various instrument and flow parameters.

With values of the flow and instrument parameters
appropriate to equipment used at Imperial College, reference [3] determines the required laser power as
0.005 mW/m/s, compared with a figure of 0.5 mW/m/s
evaluated in the earlier contribution by Durst and
Whitelaw [13] where band-pass filtering was not assumed.
Helium-neon lasers with outputs as low as 0.5 mW have

been used at Imperial College to make measurements in
flows with velocities around 1 m/s. It is undoubtedly
more convenient, however, to use lasers with output intensities closer to 5 mW.

4.2   *Optical Arrangements*
4.2.1   *General remarks*

At an early stage of the research analyses were
made (ref [1,14]) which led to the evaluation equations
for laser-Doppler anemometers.   Derivations of the
evaluation equations had, of course, been carried out
by previous users of laser-Doppler anemometers but
references [1] and [14] demonstrate clearly the relationship between optical arrangements which gave rise
to real interference fringes in the measuring control
volume and those which did not.   The implications of
the light-intensity distributions obtained with different arrangements were also revealed although a more
complete understanding of the potential advantages of
the reference-beam arrangement only emerged later from
the work of Drain [4].

It is not appropriate here to repeat the derivations referred to above but it may be useful to attempt to summarise the conclusions.   It is important
first to appreciate that the evaluation equation appropriate to laser-Doppler anemometers can be derived from
fringe, Doppler-shift or light-wave considerations and
that the optical modes represented by the diagrams of
Fig. 4 are closely related.   Indeed, the symmetrical
layout of the fringe mode of Fig. 4a and the reference-
beam arrangement of Fig. 4b represent special cases of
a more general situation in which scattered light can
be collected from any location and combined with that

from any other location.   It is clear from the equation
which represents the amplitude of the signal containing
the velocity equation,

$$i.e. \qquad I \;=\; I_1 + I_2 + 2\,\sqrt{I_1 I_2}\,\cos 2\pi \nu_D t$$

that it is desirable that the intensities of the light
waves which are to be combined are of the same magnitude.   The choice of the particular arrangement will
depend upon the purpose.   The symmetrical fringe
arrangement of Fig. 4a is convenient for many flow situations but, when the length of the measuring-control
volume is to be made as small as possible, it may be
discarded in favour of an asymmetric arrangement which
might be of the type shown in Fig. 4b;   alternatively,
it may be similar to that of Fig. 4a but with the collection arrangement at an angle to the plane of the
paper.

The reference-beam arrangement of Fig. 4b is probably to be recommended only when the number of particles
in the measuring-control volume at one time exceeds ten
or so.   As Drain [4] suggested, no advantage is gained
unless the light-collecting system is arranged to operate coherently thus causing the light waves scattered
from each of the particles in the control volume to provide in-phase signals with accumulative intensities.
Clearly, the potential advantage of this arrangement
will be limited by the loss of light intensity associated with particles outside the measuring control volume.

The one-beam arrangement of Fig. 4c is particularly
easy to understand in terms of the Doppler effect and
has performance characteristics essentially the same as

the fringe arrangement.   It offers the potential advantage, pointed out by Durst and Whitelaw [15], that
it can be used for measurements of one-point correlations of two velocity components without the need
for special polarisation effects.   This configuration
demonstrates the possibility of measurement uncertainties due to the finite range of frequencies collected
and combined;   this measurement error, frequently
termed "aperture broadening", was discussed by Durst
and Whitelaw [13] and is closely related to the transit
time errors discussed in Sec. 4.2.3.

Figure 4 also indicates the particular light-
collection systems and these emphasise the need for
arrangements which define the measuring control volume,
by means of an aperture, and which allow a magnificatio;
of the observed region on the photocathode.

4.2.2   *Dependence of signal intensity on particle size*

It can readily be imagined that the relationship
between fringe spacing and particle size will affect
the amplitude of the signal which contains the velocity
information.   A particle of diameter equal to the
fringe spacing always occupies one half cycle of dark-
ness and one half cycle of light and any movement of
the particle only rearranges the position of the light
intensity distribution over the particle cross section:
the corresponding variation of light intensity observed
by a photocathode is zero.   Durst and Whitelaw [13]
provided an analysis which simplified the required inte-
grations by assuming square-sectioned particles;   their
conclusion was that the relationship between the particle diameter and the half angle between the incident
light beams should be

$$\frac{d_p}{x_{fr}} \simeq \tfrac{1}{2}.$$

Of course, it is seldom possible to control the fringe spacing over wide limits and the particles in any flow are seldom monodispersed. However, the equation reminds us of the need to consider the relationship between, for example, the mean diameter of particles which scatter observable light and the fringe spacing in designing an optical arrangement. It is also worth noting Durst's [13,16] deduction that the signal-intensity curve, as a function of increasing particle diameter, possesses zero values and an envelope of corresponding peaks which decreases asymptotically to zero. This suggests, among other things, that particles of certain diameters will not be observed by the photocathode for a given fringe spacing and light-collection arrangement: the use of a finite discrimination level by means of the signal-processing system can, therefore, lead to what appears to be particular ranges of particle sizes.   In practice, because of the unknown shape of the particles, this method of measuring particle-size distribution can only be qualitative.

4.2.3   *Correction factors*

Although optical anemometers are usually designed to minimise possible errors, the signal observed by the photomultiplier is often not equivalent to velocity and requires the application of correction factors. The signal-processing equipment may further modify the relationship between the optical signal and the processed signal.   The topics considered here are those of transit-time broadening, gradient broadening and noise

broadening: the term "broadening" is used to indicate
that the probability density distribution of frequency
is broadened, *i.e.* the rms of frequency values is increased, for reasons not dependent upon the fluid
properties.

Transit-time broadening has been the subject of
papers by many authors, notably Lumley, George and
Kobashi [17], Greated and Durrani [18], Adrian [19],
Wang [20] and Durst and Whitelaw [13].   In each of
these papers, it was demonstrated that the fractional
broadening of the probability density distribution is
inversely proportional to the measuring control volume
or, equivalently, to the number of fringes crossed by
the light scattering particles,

$$\textit{i.e.} \qquad\qquad \frac{\sigma_a}{\nu_D} \sim \frac{1}{N_{fr}}$$

where the proportionality constant is of order unity.
In these derivations it was assumed that the fringes
were straight, *i.e.* the crossing wave fronts were plane
so that optical broadening did not occur.   As a consequence, all optical arrangements used at Imperial
College with signal processing instrumentation subject
to transit-time effect, *i.e.* spectrum analysis and frequency-tracking demodulators, were operated with more
than 100 fringes in a measuring control volume defined
by two interfering plane wave points.   Thus the required correction for transit-time broadening was always
of order 1% and, as measured by Melling and Whitelaw
[21], varied from 1.5% to 0.5% when the signal-
processing arrangements included a DISA (55L20) frequency-tracking demodulator.

An implication for an optical arrangement with 100

fringes or more is that significant spatial gradients
of velocity may occur across the measuring control volume.   This may lead to *gradient broadening*, a topic
considered in detail and with particular relation to
correcting mean velocities by Melling [22].   He recommended a correction procedure which leads to

$$\bar{U} - U_o \simeq \frac{\sigma^2}{2}\frac{\partial^2 U_o}{\partial z^2} + \frac{\sigma^4}{8}\frac{\partial^4 U_o}{\partial z^4}$$

where $\bar{U}$ is the measured velocity averaged over time
and the control volume, U_o is the true time-mean
velocity at the centre of the control volume, σ is
$d_1/2\sqrt{2}$ and z is the coordinate direction of the velocity gradient.

Vlachos [23] has also been concerned with gradient
broadening in order to determine the dimensions of
measuring control volumes.   Since finite rms levels
are measured in the laminar fully-developed flow in
small channels, it is possible to relate these to control-volume dimensions provided the velocity profile
is known.   The paths of a number of particles (100 or
1000) through the control volume were positioned by a
set of random numbers and the velocity of a particle
at a given position determined from the theoretical
profile.   Thus mean and rms values were calculated for
various values of the appropriate control volume dimension and matched to the experimental value to give the
value of the corresponding experimental control-volume
dimension.

Electronic noise is a topic which belongs to signal-processing arrangements but is mentioned here as a
reminder that the appropriate correction, together with
those for transit-time and gradient effects, should be

applied according to the equation

$$\left(\overline{\frac{e^2}{E^2}}\right)_t = \overline{\frac{e^2}{E^2}} - \left(\overline{\frac{e^2}{E^2}}\right)_a - \left(\overline{\frac{e^2}{E^2}}\right)_g - \left(\overline{\frac{e^2}{E^2}}\right)_n$$

Because of refraction, corrections to the location of measurement may also be necessary and have been discussed by Melling and Whitelaw [21].

4.3   *Signal-processing Systems*

The three methods of signal processing, discussed in Sec. 3.3, have been developed and used at Imperial College.   In each case, the low-frequency signal associated with particle presence was removed using a high-pass filter.   In highly turbulent flows, where it is difficult to locate the filter in the frequency domain, frequency shifting devices were incorporated and so minimised the influence of the filter on the Doppler signal.   The application of two of these frequency shifting devices is described in Sec. 6.

Rules for the operation of *frequency analysers* are provided by Whitelaw *et al*. [24] together with a clear indication of the principles involved.   A detailed investigation of frequency analysis for the processing of laser-anemometer signals was carried out by Asalor [25] who showed that results obtained with the square of the analyser-output signal were in close agreement with the properties of the input deterministic signal.   He extended a previous analysis of Drain [26] to confirm that, provided the filter bandwidth is infinitely thin, the analyser output signal is proportional to the square root of the probability density. The derived relationship is:

$$W(f) = \tau \, \frac{\overline{\alpha_n}}{2T} \sqrt{n_T P(\nu_D)}$$

where $W(f)$ is the frequency-analyser output signal, τ is the integration time, α_n the amplitude of the input signal and n_T the number of particles passing through the control volume in time T. In addition to confirming the relationship between $W(f)$ and $P(\nu_D)$, the equation indicates the need for a constant concentration of particles which scatter light with constant amplitude.

In order to avoid effects of a finite-filter bandwidth, Asalor recommended that the ratio of spectrum width to filter width should exceed 15. In addition, the scan rate and filter bandwidth should satisfy the condition that:

$$\frac{df}{dt} \leq 5\Delta f_o$$

where Δf_o is the bandwidth. If these and other recommendations are followed, it is possible to use a frequency analyser to effect measurements of mean and rms values of frequency with precisions of the order of ±1.5% and ±4% respectively.

Contributions to the understanding of *frequency-tracking demodulation* are of a similar type to those for frequency analysis. They were required to construct the phase-locked system presently in use at Imperial College; but the more important contributions have probably been to the testing and evaluation of the existing instruments manufactured by DISA and Cambridge Consultants Ltd. rather than to the introduction of new principles to the Imperial College tracker. Principles

of operation and rules for the use of frequency-tracking
demodulation are again provided by Whitelaw *et al.* [24]
and comments specific to the two instruments mentioned
above are contained in the reports by Durão and Whitelaw
[27,28].   An important conclusion from the tests carried out with commercial trackers is that the availability of a signal mixer, which allows the centre frequency of a given signal to be adjusted to a preferred
location on a given frequency range, can significantly
improve frequency response.

Counting techniques have been used in several investigations, for example those of Baker, Hutchinson
and Whitelaw [9,10] and Durão and Whitelaw [29];   these
have shown the advantages of high quality signals of
logic circuitry to avoid erroneous counts, of carefully
matched filters and of light-frequency shifting.

High quality signals are, of course, obtained from
well designed and arranged optical systems and permit
zero-crossing counting which, in turn, allows one to
avoid ambiguity errors arising from counting periods at
finite discrimination levels.   This topic is discussed
further by Baker, Hutchinson and Whitelaw [10].   Various
forms of logic circuitry can be used to avoid erroneous
counts due to counting from one signal envelope to the
next;   the problem is not severe if light-frequency
shifting is employed.   The use of light-frequency shifting arrangements increases the number of cycles in any
signal envelope according to the formula:

$$n = (\nu_s + \nu_D) \frac{d}{U}$$

where   d   is the control-volume diameter,   U   the particle velocity, ν_D the Doppler-frequency shift and ν_s

the light-frequency shift.   At high turbulence intensities, very high frequency variations make it difficult to separate the so-called pedestal signal from the
low-frequency Doppler signals.   In addition, real filter characteristics combined with a preset-counter-
trigger level tend to suppress the tail of the probability distribution.   Light-frequency shifting mini-
mises these potential error sources by reducing the
relative range of the frequency variations as indicated
by the above equation.

5.   CONTRIBUTIONS TO EQUIPMENT

Our early investigations into the physics of optical anemometry made use of commercially available and
individual optical components: work of this type has
been described by Durst and Whitelaw [1].   Optical
arrangements of this type proved adequate for one-
dimensional velocity measurements under controlled
laboratory conditions and, indeed, are still used to
allow us to improve understanding and to develop new
systems.   However, unacceptable re-positioning errors
were experienced with these non-integrated optical
arrangements and this, together with the need for
readily transportable and prealigned systems, led to
the development of integrated systems.

Four basic types of integrated optical arrangements
have been built and are in current use.   The first has
been described by Durst and Whitelaw [15] and possesses
the following features:

- it permits rapid and precise alignment,
- it allows measurements of two orthogonal velocity components by rotation of the beam-
  splitter/mirror arrangement,

Fig. 8

- it is robust and can be used in hostile environments,
- it can be used in fringe, reference-beam or
  single-beam modes,
- it permits rapid change of focussing lens.

This arrangement has proved to be simple but versatile
and appears to have formed the basis for the DISA optical system.   The second integrated arrangement em-
bodied two beam-splitter/mirror arrangements and so
could measure, on-line, the correlation between fluctuating velocities in two orthogonal directions, *i.e.*
the Reynolds shear stresses.

The above optical arrangements have been used exclusively with low power lasers.   With larger-power
lasers, the finite effective coherence associated with
multi-mode operation dictates that the path lengths of
the two light beams should be equal.   This requirement

Fig. 9

is fulfilled in the two-channel integrated optical
arrangement described by Durst and Whitelaw [3].
Light-path compensators are incorporated in each of
the two orthogonal channels of this equipment.

Figure 8 shows a photograph of each of the three
optical arrangements discussed in the previous paragraphs.   In each case, the optical arrangement is
mounted in a holder which allows three orthogonal trans-
lational adjustments and two rotational adjustments.
The holder itself is fixed to a base which forms part
of an optical bench.   An example of a complete optical
bench arrangement comprising laser, optical arrangement,
light collecting lens and photomultiplier with aperture
is shown in Fig. 9.   Arrangements of this type have
greatly facilitated our measurement programme.

The optical arrangements shown in Fig. 8 *can* be

Fig. 10
operated with backward scattered light but are not
ideal for this purpose.   In addition, they are made
up of beam-splitter/mirror assemblies which can be adjusted with high precision: this was necessary at the
time of their design because suitable prisms could not
be obtained with sufficiently close tolerances.   More
recently, prisms with the required precision have become available and have been incorporated in the optical arrangement shown in Fig. 10.   This unit was designed with considerable help from Dr. F. Durst, now
of the Sonderforschungsbereich 80 at the University of
Karlsrühe, and is intended for measurements in these
situations where operation with forward scattering light
is inconvenient or impossible.

        These optical arrangements, together with the more
specialised arrangements such as that used by Baker,

Hutchinson and Whitelaw [30], have been designed in
accord with the principles outlined in Sections 4.1
and 4.2.

A rotating grating, a Bragg cell and a Kerr cell
have, on different occasions been incorporated in optical arrangements to allow light-frequency shifting.
Rotating gratings are well known (see for example Baker
[31] and are readily available provided an efficiency
of 20% or less is acceptable: higher efficiencies can
be achieved and such gratings are presently under examination. The Bragg cell has been in operation for
some months and has made use of crystals with oscil-
lation frequencies as low as 4 MHz; its use is referred to by Durão and Whitelaw [29] and design details
have been provided in their more recent report [32].
It is worth noting that a 9.2 MHz crystal can be used
to result in zero, plus and minus first-order shifted
beams with 90% of the light intensity distributed evenly
between them. Alternatively, at its third harmonic,
zero and a first-order beam can be obtained with 40%
of the intensity in each. A Kerr cell was used by
Baker, Hutchinson and Whitelaw [9] and was kindly made
available by Dr. L.E. Drain; the instrument has been
described by Drain and Moss [7].

Very recently, Baker, Hutchinson, Khalil and Whitelaw [33] have reported measurements obtained using a
rotating grating and a filter bank signal processor.
This instrument was designed at AERE, Harwell and shows
great promise for measurements in turbulent flows.

Additional contributions have been made to the design and construction of equipment for the generation
of seeding particles: some of these are described by
Durst and Whitelaw [34], Zaré [35] and by Melling and

Whitelaw [12].

6.   SOME APPLICATIONS OF LASER-DOPPLER ANEMOMETRY

     The present section has been divided into three
sub-sections which relate to measurements in water, air
and combusting flows respectively.   The purpose of the
section is to indicate the various flow configurations
which have been investigated with laser-Doppler anemometry equipment by reference to publications and to
describe a small number of investigations in sufficient
detail to allow the state of the art and the remaining
difficulties to be identified.

6.1   Water Flows

     Measurements in water flows have been described
by Durst and Whitelaw [15], Briard [36], Bhatia [37],
Crane and Melling [38], Gosman, Vlachos and Whitelaw
[39] and by Melling and Whitelaw [21].   Those of Durst
and Whitelaw demonstrated that laser-Doppler anemometry
equipment resulted in values of mean velocity and longitudinal normal stress in fully-developed, turbulent
channel flow which were in close agreement with the accepted hot-wire measurements of Comte-Bellot [40].
Briard investigated the possibility of applying a laser-
Doppler anemometer to vapour-liquid bubbly flows but
the results were inconclusive.   Bhatia made measurement
of mean velocity and longitudinal normal stress in a
90° bend of a square duct channel: the initial conditio
for these measurements was, however, unsatisfactory and
the measurements are presently being repeated.   Gosman,
Vlachos and Whitelaw were concerned with the laminar flo
in a round tube downstream of a 15° bend and presented

measurements which demonstrated the growth of the in-
itially skewed velocity profile downstream of the bend
to a fully-developed parabolic profile; the measurements agreed closely with values obtained from a numerical solution of the appropriate differential
equations. Related work by Vlachos and Whitelaw [41]
has shown the extent to which similar equipment can be
used for the measurement of blood velocity. The remaining investigations in water flows deserve particular mention and relevant comments are provided below.

The brief experimental program reported by Crane
and Melling, required the application of laser-Doppler
anemometry to the two-phase, wet-steam flow downstream
of a small Curtis-type turbine. The work was performed to find out the problems associated with the
measurement of velocity in wet-steam flows and to quantify the range of dryness fractions for which measurements were possible. The original intention was to
compare measurements of mean velocity made by laser-
Doppler anemometry with those obtained from an impact
probe, so determining the precision with which impact
probes could measure. The investigation was undertaken at the Central Electricity Research Laboratories,
Leatherhead with the assistance of staff of the Laboratories.

The optical anemometer consisted of an argon-ion
laser operated at powers up to 250 mW, an integrated
optical unit operated in forward-scatter, fringe mode
with a 300 mm focal length lens. Signal processing
was by means of a Hewlett Packard spectrum analyser
(Model 8552A/8553B). The steam tunnel was 152 mm wide
by 305 mm high and had Schlieren glass windows through
which the laser light was transmitted and received. A

range of test-section pressures from 0.05 to 0.28 bar,
of velocities from 50 to 200 m/s and of dryness fractions from saturation to 95% was investigated.   Calculations, based on a number concentration of 0.5 μm
diameter particles suggested that several hundred droplets were simultaneously in the measuring volume.
Observation of the photomultiplier signal suggested
that the concentration was of the order of one hundred
but perhaps slightly lower.   This high concentration
resulted in poor quality signals which were not improved
by attempts to make use of a reference-beam system.
Nevertheless, the signals were resolvable by frequency
analysis.

The measurements indicated that, for values of dryness fraction in excess of 0.98, precise measurements
of mean velocity could be obtained.   Because of the
low pressure and multi-particle-broadening effects,
high-frequency resolution of the vapour motion could
not, however, be achieved.   The measured values of
mean velocity agreed with impact-probe results to within
2.3%.

The more extensive investigation of reference [21]
had, as its main purpose, the provision of measurements
of developing non-circular duct flow, particularly with
a view to providing a severe test for numerical calculation procedures such as those described by Dr. Patankar
in the first article of this volume.   The measurements
were obtained in a rectangular duct of dimensions 40 mm
× 41 mm and 1.8 m long.   Detailed contour plots of
axail mean velocity and the corresponding normal stress
were obtained at five downstream stations between 5.6
and 37 hydraulic diameters from the duct entrance at a
Reynolds number of 3.9×10^4.   Subsequently, Melling

[42] extended these measurements to include values of
all six components of the Reynolds stress at the first
and last measuring stations.

These measurements were obtained with a 5 mW laser,
a two-channel integrated optical unit and signal processing by frequency-tracking demodulation (DISA 55L20).
The 9 mm thick plexiglass presented no serious problems
and the high quality signal, with drop-out of less than
2%, obtained from the dirt particles in water allowed
detailed and precise measurements. The unfiltered
signal indicated a ratio of maximum amplitude of Doppler
burst to the average amplitude of the low-frequency
signal associated with particle presence of about 0.2;
band-pass filtering from 100 kHz to 2 MHz was, however,
used for the measurements.

Consideration of the corrections to the results
obtained from the frequency tracker indicated that their
maximum values were of the order of 3% of the rms values.
Since the influence of transit-time, gradient and noise
broadening was known with reasonable certainty, the
corrected measurements were more precise than obtained
previously by hot-wire methods.

The rectangular-duct water flow is representative
of a large class of flows which are particularly suited
to investigation by optical anemometry instrumentation.
The use of water and comparatively small dimensions allows the use of compact anemometry equipment with frequency tracking demodulation: the absence of turbulence
intensities in excess of 10% makes frequency tracking
particularly convenient and obviated the need for light-
frequency shifting. Because of this convenience and
the associated precision and speed of measurement two
further investigations, which otherwise might have used

air flow, are now in progress at Imperial College using
water flow.

Further development work is undoubtedly necessary
to extend the applicability of optical anemometry to
a useful range of two-phase flows of the liquid-vapour
type.   The experiments of Crane and Melling [38],
Briard [36] and Durst and Whitelaw [34] demonstrate
some successes but these are only on the fringe of the
two-phase flows of interest to process engineers. Our
present intention is to attempt to use a combination
of forward and backward scattered light as suggested
by Davies [43] and Davies and Unger [44] to extend this
range of application.

6.2   *Air flows*

Air flows have been investigated by Durst and
Whitelaw [13,34,45], Durst, Melling and Whitelaw [46],
Melling [47], Durão and Whitelaw [27,28,29], Baker
[31,48], Baker, Hutchinson, Khalil and Whitelaw [33].
The more recent investigations are characterised by a
much greater emphasis on results of significance to
fluid mechanics although, as the flow complexity and
required precision increases, further instrument developments have proved necessary. This trend is readily demonstrated by the various measurements in free-jet flows.
Durst and Whitelaw [45] presented measured values
of mean velocity, one normal stress and shear stress in
a turbulent-air jet.   The measurements were obtained
with a single-channel laser-Doppler anemometer and a
prototype frequency-tracking demodulator.   They were
limited to a region within 8 diameters of the jet-exit
because the increase in turbulence intensity and decrease in particle concentration with downstream dis-

tance prevented the successful operation of the tracker beyond this region.

Some two years later, Durão and Whitelaw [27] made measurements in a turbulent air jet with a similar optical arrangement to that of reference [45] but with a new frequency-tracking demodulator. On this occasion measurements of mean velocity, three normal stresses and shear stress were obtained to downstream distances in excess of 25 diameters: the highest rms turbulence intensity recorded was of the order of 25% of the mean velocity.

Although these measurements covered a much larger region of the flow they still did not extend to values of the radial distance greater than approximately 0.15 times the downstream distance.   In the region close to the edge of a jet, the turbulence intensity exceeds 30% and the probability of negative velocities becomes increasingly more significant.   In addition, if the jet is seeded, the particle concentration decreases towards the edge.   The former difficulty is overcome by introducing light-frequency shifting; Baker [31, 48] and Durão and Whitelaw [29,32] have adopted this procedure.

Because of the need for instrumentation which will operate with widely different values of particle concentration*, frequency-tracking demodulation has now been replaced by filterbank and counting procedures. Baker [31] used an optical arrangement based on a grating, with 10,800 lines, which was rotated at

* Our experience of the DISA frequency tracking demodulator (Model 55L20) and our own phase-locked loop instrument indicates that they operate satisfactorily only when the particle concentration is high and constant with time: this finding may not be true of other instruments.

3000 rpm by a synchronous motor; the resulting shift
in the first-order diffraction pattern was 1.08 MHz.
The filter bank used for signal processing and described by Baker [49] had 50 filters in the fequency
range from 0.631 to 6.025 MHz.   With the 1.08 MHz
frequency shift, the measurable velocity range was
approximately - 2 m/s to 23 m/s.   Baker presented extensive measurements in his turbulent-jet flow.   The
measurements included mean velocity, three normal
stresses, Reynolds shear stress and the skewness, flat-
ness and probability functions for the streamwise velocity component; the centre-line measurements extended
to 79 diameters and the radial measurements to 0.2
times the downstream distance.   No seeding was used
and the argon-ion laser was operated with a power output of approximately 200 mW.

    Durão and Whitelaw [29] investigated the flow
downstream of an annular jet using a Bragg cell and
counting techniques; the flow geometry caused a recirculation zone on the axis immediately downstream of
discharge.   This cold-flow investigation provided a
comparison for subsequent measurements in the same
geometrical configuration but with a flame of town gas.
Although the recirculation zone was of greatest interest
detailed measurements were obtained in the downstream
region of the jet.   These data together with the
measurements of Baker, constitute a more extensive investigation of free-jet flows than has previously been
reported in the literature.   No seeding was required
for this investigation since an Argon-ion laser was
used.   The Bragg cell provided a frequency shift of
8.7 MHz and considerable care had to be exercised in
the use of high- and low-pass filters in order to en-

sure that turbulence energy was not accidentally fil-
tered out. (Baker [31] has also provided comments
relating to the selection of filters for use with
counting procedures.)   The increased number of fringes
resulting from the 8.7 MHz shift made the operation
of the counter comparatively easy since the risk of
counting from one signal envelope to another became
much less likely and simple logic sufficed.

Although the argon-ion laser allowed measurements
without seeding, this was not possible in those investigations which employed a low power laser.   In free
flows, fluid is entrained from the surroundings and
this, necessarily unseeded fluid, is not recognised by
an anemometer designed to follow seeding particles.
The resulting signal will therefore have a probability
distribution which is biased towards the particles
originating in the core flow.   The magnitude of possible errors was investigated in references [48] and
[50].   The latter measurements were obtained in a free-
jet flow and the error shown to be negligible except
for radial positions greater than 0.15 times the downstream distance.   The maximum recorded effect on mean
velocity was 3% of the centre-line value and occurred
at the measuring station closer to the jet exit, *i.e.*
15 diameters downstream, and at a radial position of
0.2 times the downstream distance;   the corresponding
effect on the rms value was approximately ten times
greater.   Although these errors are negligible for
many practical purposes, it is important to quantify
possible errors due to particles and their origin.
For this reason, the preliminary investigation is at
present being extended to investigate the influence of
greater relative particle concentrations in jet flows

and in simple diffusion-flame configurations. The
latter may prove to be particularly important if the
suggestions of Durst and Zaré [51] regarding the os-
cillation of the reaction zone with respect to the
measuring volume prove to be well founded

Durst, Melling and Whitelaw [46] have demonstrated
that, just as in water flows, laser-Doppler anemometry
can conveniently be used to obtain precise measurements
in flow configurations of small dimensions. Under
these circumstances, controlled seeding can be employed
and allows the use of frequency-tracking demodulation.
Some frequency-tracking demodulators may also be used
in free flows although light-frequency shifting is
necessary in regions where the turbulence intensity exceeds about 25%. A filter bank or counter may also
be convenient in air flows: the former offers particular advantages in highly turbulent flows where the
signal quality may be poor. In addition, since the
filter bank measures particle velocity and weighs each
velocity value according to its duration, it does not
suffer from the bias effects presently under examination
in connection with counting systems. We expect these
bias effects to be significantly smaller than suggested
by the analysis of McLauchlan and Tiederman [52] but
this remains to be proven. In both cases, light-
frequency shifting is desirable where the turbulence
intensity exceeds 20% and is necessary above 30%.

6.3 Combusting Flows

Laser-Doppler anemometry offers the possibility
of measuring velocity and its correlations in combusting
flows where such measurements were previously impossible
Because of the importance of combusting flows, we have

made many measurements in a variety of combustion configurations.   Durst and Whitelaw [13], Asalor [53],
Zaré [35], Durst, Melling and Whitelaw [54,55,56],
Baker, Bourke and Whitelaw [57,58], Baker, Hutchinson
and Whitelaw [9,10,30], Durão, Melling, Pope and
Whitelaw [59], Durão and Whitelaw [60], Durão, Durst
and Whitelaw [61], Baker, Hutchinson, Khalil and
Whitelaw [33] describe some of these measurements.
The range of flow configurations investigated and described in these papers is large and further comments,
relating particularly to the range of geometrical size,
are provided below.

Most of our investigations have been carried out
under laboratory conditions where all instrumentation
was readily available and controlled seeding could be
employed.   For example, the investigation of laminar,
oscillating flames conducted by Durão and Whitelaw
[60] allowed controlled seeding and, therefore, the
use of a frequency-tracking demodulator: the laboratory nature of the investigation also allowed the
measurement of mean temperature, rms temperature fluctuations and the velocity-temperature correlation.
The signal was analysed to determine precisely the frequency of the flow oscillations.

The measurements described by Baker, Hutchinson
and Whitelaw [9,10] were also carried out in the laboratory but were concerned with a larger flame stabilised
on an industrial burner.   Once again, controlled seeding was possible but the high turbulence intensities
and relatively small proportion of time for which the
signal was present, required the use of a counter.
Light-frequency shifting was also employed to allow
measurements in and around regions of recirculation and

made use of an electro-optic modulator consisting of
a 4-plate Kerr cell operated at 3.25 MHz and manufactured by colleagues at Harwell.   As in the case of
other laboratory investigations, few difficulties were
experienced once the correct instrumentation had been
assembled.   The resulting measurements are novel and
could not readily have been obtained by other means.

An attempt to measure in a 2 m square furnace is
reported by Baker, Hutchinson and Whitelaw [30] and
reveals some of the practical problems which may be
encountered in such flows.   No measurements were possible with coal and gas flames in the region near the
burner exit.   The coal flame contained too many solid
particles and attenuated the laser light to an extent
which prevented a measurable signal.   In contrast,
the gas flame contained too few naturally available
particles to allow statistically meaningful measurements within a reasonable period of time.   Measurements were possible in an oil flame but in a region
where liquid droplets probably contributed to the signal.   This investigation was conducted using a
specially designed, forward-scattering optical arrangement, an argon-ion laser operating at 200 mW and a
frequency analyser.

In addition to the particle problem (which may be
overcome by local seeding) the 2 m-furnace investigation demonstrated that the collimation of a laser
beam is partially destroyed by the refractive-index
fluctuations present in combustion systems.   In addition the laser beam is moved, in space, by these
variations in refractive index;   as a result, the
Doppler signal is present less often than it would be
in the absence of refractive-index gradients.   The

combination of the diminution in light intensity at
the region of beam intersection due to the loss of
collimation and the relative movement of the two light
beams and the collected light, means that the possibility of measurement in combusting flows is likely to
diminish rapidly with scale and may be impossible for
scales not much larger than 2 m.   This problem may be
overcome by arranging for the light path to traverse
only a small distance of the hot-flow and will certainly be reduced by the use of instrumentation such
as the filterbank: these possibilities remain to be
quantitatively evaluated.

7.   CLOSURE

The survey of work undertaken by my colleagues
and I has attempted to show that laser-Doppler anemometry is now a well-developed technique and can be
used for fluid-dynamic purposes with confidence and
precision.   Some questions remain to be answered but,
for the most part, these are concerned with refining
the precision of an already established technique.
The next few years will increasingly see greater emphasis on applying the existing technology to obtain
detailed measurements in important complicated flows
where hitherto experimental knowledge has been frag-
mentary or absent.
Small scale, recirculating, three-dimensional and
combusting flows can all be examined, with advantage
using laser-Doppler anemometry and, thus, our knowledge
of the nature of the flow in venules, in tube banks
and other form of heat exchangers, in rotating machinery,
in combustion chambers and in many other items of engin-

eering hardware is likely to be significantly improved in the near future.

Having mentioned future work, it is appropriate that I should close by acknowledging the considerable financial support which we have received. The Science Research Council has provided continuous funding since 1969 and its grants have been generously supplemented by others from the CEGB, AERE Harwell, NATO and British Heart Foundation; in addition, considerable resources of the Department of Mechanical Engineering have been made available. To all who have thus sponsored the advancement of the work I would like to express my gratitude. Finally, to the past and present members of the group, whose names appear in the cited references, I send my grateful thanks not only for their technical contributions but also for their willingness to place the good of the research team as a whole above the immediate inclinations of the individual; their efforts have led to the success of the research programme described in this review.

8.   REFERENCES

1.   Durst, F. and Whitelaw, J.H. "Optimisation of optical anemometers". *Proc. Roy. Soc.* A 324, 157, 1971.

2.   Durst, F., Melling, A. and Whitelaw, J.H. "Laser anemometry: a report on Euromech 36". *J. Fluid Mech.* 56, 143, 1972.

3.   Durst, F. and Whitelaw, J.H. "Light source and geometrical requirements for the optimisation of optical anemometry signals". *Opto-Electronic* 5, 137, 1973.

4.   Drain, L.E. "Coherent and non-coherent methods in Doppler optical beat velocity measurement". *J. Phys.* D5, 481, 1972.

5. Denison, E.B. and Stevenson, W.H. "Oscillating flow measurements with a directionally sensitive laser velocimeter". *Rev. Sci. Inst.* <u>41</u>, 1475, 1970.

6. Briard, P. and Denham, M.K. "The design and application of a directionally sensitive laser anemometer". University of Exeter, Dept. of Eng. Science Report, 1972.

7. Drain, L.E. and Moss, B.C. "The frequency shifting of laser light by electro-optic techniques". *Opto-electronics* <u>4</u>, 429, 1972.

8. Hiller, W.J. and Meier, G.E.A. "The scattered light-beam method". *Proc. of Electro-optic Systems in Flow Measurement.* University of Southampton, 1972.

9. Baker, R.J., Hutchinson, P. and Whitelaw, J.H. "Detailed measurements in the recirculation region of an industrial burner by laser anemometry". *Proceedings of the European Combustion Symposium*, 583, 1973.

10. Baker, R.J., Hutchinson, P. and Whitelaw, J.H. "Velocity measurements in the recirculation zone region of an industrial burner flame by laser anemometry with light frequency shifting". *AERE*-R7492, 1973. *Combustion and Flame* <u>23</u>, 57, 1974.

11. Melling, A. "Scattering particles for laser anemometry in air: selection criteria and their realisation". Imperial College, Dept. of Mech. Eng. Report ET/TN/B/7.

12. Melling, A. and Whitelaw, J.H. "Seeding of gas flows for laser anemometry". *DISA Information* <u>15</u>, 5, 1973.

13. Durst, F. and Whitelaw, J.H. "Theoretical considerations of significance to the design of optical anemometers". *ASME* Paper 72-HT-7, 1972.

14. Durst, F. and Whitelaw, J.H. "Aerodynamic properties of separated gas flows: existing measurement techniques and a new optical geometry for the laser-Doppler anemometer". *Progress in Heat and Mass Transfer* <u>4</u>, 311, 1971.

15. Durst, F. and Whitelaw, J.H. "Integrated optical units for laser anemometry". *J. Phys.* <u>E4</u>, 804, 1971.

16. Durst, F. "Development and application of optical anemometers". Ph.D. Thesis, University of London, 1972.

17. Lumley, J.L., George, W.K. and Kobashi, Y. "The influence of ambiguity and noise at the measurement of turbulent spectra by Doppler scattering". *Proc. 1st Biennial Symposium on Turbulence in Liquids*, Rolla, 1969.

18. Greated, C. and Durrani, T.S. "Signal analysis for laser velocimeter measurements". *J. Phys.* E4, 24, 1971.

19. Adrian, R. "Statistics of laser Doppler velocimeter signals: frequency measurement". *J. Phys.* E5, 91, 1972.

20. Wang, C.P. "Effect of Doppler ambiguity on the measurement of turbulent spectra by laser-Doppler velocimeter". *Appl. Phys. Lett.* 22, 154, 1973.

21. Melling, A. and Whitelaw, J.H. "Measurements in turbulent water flow by laser anemometry". Imperial College, Mech. Eng. Dept. Report HTS/73/44, 1973. To be published in the *Proceedings of the 3rd Biennial Symposium on Turbulence in Liquids*.

22. Melling, A. "The influence of velocity gradient broadening on mean and rms velocities measured by laser anemometry". Imperial College, Mech. Eng. Dept. Report HTS/73/37, 1973.

23. Vlachos, N. Private communication.

24. Whitelaw, J.H., Baker, R.J., Drain, L.E., Durst, F. and Melling, A. Lectures for a Post-Experience Course entitled "Optical Beam Methods for Velocity Measurements". Imperial College, Mech. Eng. Dept. Report HTS/73/7, 1973.

25. Asalor, J. "An examination of spectrum analysis for the processing of laser-anemometry signals". Imperial College, Mech. Eng. Dept. Report HTS/73/46, 1973.

26. Drain, L.E. Private communication.

27. Durão, D.F.G. and Whitelaw, J.H. "Some performance characteristics of the Cambridge Consultants frequency-tracking demodulator". Report

to Survey and General Instrument Co., 1973. (Copies available from authors.)

28. Durão, D.F.G. and Whitelaw, J.H. "Some performance characteristics of a prototype of the DISA frequency-tracking demodulator". Imperial College, Mech. Eng. Dept. Report HTS/74/11, 1974.

29. Durão, D.F.G. and Whitelaw, J.H. "Measurements in the region of recirculation behind a disc". Imperial College, Mech. Eng. Dept. Report HTS/74/14, 1974. To be published in the *Proceedings of the Laser Velocimetry Workshop, Purdue University, 1974*.

30. Baker, R.J., Hutchinson, P. and Whitelaw, J.H. "Preliminary measurements of instantaneous velocity in a two-metre square furnace using laser anemometry". *ASME J. Heat Transfer* **96**, 410, 1974.

31. Baker, R.J. "Measurements in an isothermal jet - comparison between hot-wire and laser anemometry". *AERE*-R7648, 1974.

32. Durão, D.F.G. and Whitelaw, J.H. "The performance of acousto-optical cells for laser-Doppler anemometry". Imperial College, Mech. Eng. Dept. Report HTS/74/24, 1974.

33. Baker, R.J., Hutchinson, P., Khalil, E. and Whitelaw, J.H. "Measurements of three velocity components in a model furnace with and without combustion", 1974. To be published in the *Proceedings of the 15th Combustion Symposium*.

34. Durst, F. and Whitelaw, J.H. "Local velocity measurements in atomised sprays". *Jahrbuch der DFVLR*, 188, 1971.

35. Zaré, M. "Investigation of combustion instabilities in open flames". M.Sc. Thesis, University of London, 1972.

36. Briard, P. "Laser anemometry in single and two-phase flows". M.Sc. Thesis, University of London, 1971.

37. Bhatia, H.R. "Optical anemometry measurements in the 90° bend of a square-flow duct". M.Sc. Thesis, University of London, 1972.

38. Crane, R.I. and Melling, A. "Velocity measure-
    ments in wet steam flows by laser anemometry
    and pitot tube". Central Electricity Research
    Laboratories Report, RD/L/N158/73, 1973.

39. Gosman, A.D., Vlachos, N. and Whitelaw, J.H.
    "Laminar pipe flow downstream of a 15° bend".
    Imperial College, Mech. Eng. Dept. Report
    HTS/73/4, 1973.

40. Comte-Bellot, G. "Écoulement turbulent entre
    deux parois parallèles". *Publ. Sci. et Tech.
    du Ministère de l'Air*, 1965.

41. Vlachos, N. and Whitelaw, J.H. "The measurement
    of blood velocity with laser anemometry".
    *Proceedings of the Laser Velocimetry Workshop,
    Purdue University* 1, 521, 1975.

42. Melling, A. Ph.D. Thesis, University of London,
    1974.

43. Davies, W.E.R. "Velocity measurement in bubbly
    two-phase flows using laser-Doppler anemometry"
    University of Toronto Report UT/AS TN184, 1973.

44. Davies, W.E.R. and Unger, J.I. "Velocity measure-
    ments in bubbly two-phase flows using laser
    Doppler anemometry". University of Toronto
    Report UT/AS TN185, 1973.

45. Durst, F. and Whitelaw, J.H. "Measurements of
    mean velocity, fluctuating velocity and shear
    stress using a single channel anemometer".
    DISA Information 12, 11, 1971.

46. Durst, F., Melling, A. and Whitelaw, J.H. "Low
    Reynolds number flow over a plane symmetrical
    sudden expansion". *J. Fluid Mech.* 64, 111,
    1974.

47. Melling, A. "The laser-Doppler shift technique
    and hot-wire anemometry: a comparison".
    M.Sc. Thesis, University of London, 1970.

48. Baker, R.J. "The influence of particle seeding
    on laser anemometry measurements". *AERE*-M2644,
    1974.

49. Baker, R.J. "A filter bank signal processor for
    laser anemometry". *AERE*-R7652, 1974.

50. Durão, D.F.G. and Whitelaw, J.H. "Performance

characteristics of two frequency-tracking demodulators and a counting system: measurements
in an air jet". *Proceedings of the Laser Velocimetry Workshop, Purdue University* 1, 170,
1975.

51. Durst, F. and Zaré, M. "Velocity measurements
in turbulent premixed flames by means of laser
Doppler anemometers". Report of Sonder-
forchungsbereich 80, University of Karlsrühe
SFB/EM/3, 1973.

52. McLauchlan, D.K. and Tiederman, W.G. "Statistical biasing in individual realisation laser
anemometry". Oklahoma State University
School of Med. and Aero. Eng., Report ER-73-
J-19, 1973.

53. Asalor, J. "Investigation of combustion insta-
bility in confined flames". M.Sc. Thesis,
University of London, 1972.

54. Durst, F., Melling, A. and Whitelaw, J.H. "The
application of optical anemometry to measurement in combustion systems". *Combustion and
Flame* 18, 197, 1972.

55. Durst, F., Melling, A. and Whitelaw, J.H. "Optical anemometer measurements in recirculating
flows and flames". *Proc. DISA Conference:
Fluid Dynamic Measurements in Industrial and
Medical Environments*, 81, 1972.

56. Durst, F., Melling, A. and Whitelaw, J.H. "Laser
anemometry measurements in a square duct with
and without combustion oscillations". Imperial
College, Mech. Eng. Dept. Report EHT/TN/A/40,
1972.

57. Baker, R.J., Bourke, P.J. and Whitelaw, J.H.
"The application of laser anemometry to the
measurements of flow properties in industrial
burner flames". *14th Combustion Symposium*,
699, 1973.

58. Baker, R.J., Bourke, P.J. and Whitelaw, J.H.
"Measurements of instantaneous velocity in
laminar and turbulent diffusion flames using
an optical anemometer". *J. Inst. Fuel*, 388,
1973.

59. Durão, D.F.G., Melling, A., Pope, S.J. and

Whitelaw, J.H. "Laser anemometry measurements in the vicinity of a gutter-stabilised flame". Imperial College, Mech. Eng. Dept. Report EHT/TN/A/41, 1972.

60. Durão, D.F.G. and Whitelaw, J.H. "Instantaneous velocity and temperature measurements in oscillating diffusion flames". *Proc. Roy. Soc.* A338, 479, 1974.

61. Durão, D.F.G., Durst, F. and Whitelaw, J.H. "Optical measurements in pulsating flames". *J. Heat Transfer* 95, 177, 1973.

9.   NOMENCLATURE

Symbol	*Meaning*
d_p	particle diameter
d_1	diameters of control volume at $1/e^2$ - intensity locations
e	fluctuating voltage
E	voltage
E(U)	energy
Δf_o	filter bandwidth
h	Planck's constant
I	intensity
n_T	number of particles in time T
N_{fr}	number of fringes
N_{ph}	number of fringes observed at photo-multiplier
P	probability
P	laser power
Q_{scat}	scattering coefficient
t	time
$u_{1,2,3}$	fluctuating component of velocity
$U_{1,2,3}$	velocity
W	frequency-analyser output signal

Symbol	Meaning
x_{fr}	fringe spacing
α_n	amplitude
η_c	efficiency of light collection system
η_q	quantum efficiency of photocathode
λ	wave length of light in air
ν_D	Doppler frequency
ν_o	frequency of laser light
ν_s	frequency of shifted light
σ_a	rms width of Doppler spectrum due to finite lifetime of signal
τ	integration time
ϕ	half angle between light beams in air

Subscript	Pertaining to
A, B	spatial locations
a	relating to transit time
g	relating to velocity gradient
i, j, k	tensor subscripts which may take the values 1, 2 and 3
n	relating to noise
o	relating to centre of scattering volume
t	relating to turbulence

Superscripts	Pertaining to
$\wedge$	instantaneous value
$-$	time average value
$\sim$	rms value

THE BEHAVIOUR OF TRANSPIRED TURBULENT
BOUNDARY LAYERS

by

W.M. Kays and R.J. Moffat

Department of Mechanical Engineering, Stanford University
Stanford, California, USA

ABSTRACT

The article collects together the main experimental results on transpired turbulent boundary layers
obtained by the authors and their colleagues at
Stanford University over the nine-year period 1965-74.
The data span measurements of both the thermal and
momentum boundary layer development for both injection
and suction.   The simultaneous effects of accelerating
or decelerating the mainstream flow are also explored.
in some detail.

Particular emphasis has been placed on equilibrium
boundary layers, that is flows where the rates of transpiration and variations of free-stream velocity along
the test plate are adjusted so that the boundary layer
structure remains virtually the same as the flow develops.   In this way, effects of the transpiration
and pressure gradient may be studied free from the
masking effect of upstream history.

The observed flow behaviour is shown to be well
accounted for with a version of Prandtl's mixing-length
hypothesis, provided the influence of transpiration and
pressure gradient on the viscous sub-layer thickness
are included.

1.   INTRODUCTION

The boundary layer with transpiration through the
solid surface is a variant of the general boundary laye
problem that has been of considerable interest in technical applications for at least two decades.   In the
early 1950's transpiration was being extensively investigated as a means of cooling aerodynamic surfaces unde
high velocity flight conditions.   But transpiration
from a solid surface over which a fluid is flowing, and
on which a boundary layer is developing, is of interest
in a large number of quite different types of applications, of which transpiration cooling is only one.

In the typical transpiration cooling application,
the solid surface is constructed of some kind of porous
solid material.   Cooling fluid, which may be chemicall
the same as the free-stream, is then forced through the
surface with the objective of protecting the surface
from a hot free-stream.   This is a boundary-layer prob
lem for which the normal component of velocity at the
solid-fluid interface is non-zero, but otherwise the
same momentum and energy boundary-layer differential
equations must be solved as for the non-transpired
boundary layer.   A variation on this problem occurs
when the cooling fluid is a chemically different specie
from the free-stream fluid.   For example, helium might
be injected as a coolant to protect a surface from a
high temperature air stream.   In this case the mass-
diffusion equation of the boundary layer must be solved
in addition to the momentum and energy equations.   There
are obviously similarities between these two types of
problems, but also fundamental differences.   Both are
"mass-transfer" problems in the sense that mass is trans

ferred across the fluid-solid interface, but the latter
is also a mass-diffusion problem, while the former is
not.

Another transpiration problem arises when there
is evaporation or sublimation from an interface into a
boundary layer, or condensation onto the interface.
A further variation on the problem arises when there
is chemical reaction either within the boundary layer
or at the surface.

In any of the cases cited, the direction of the
flow normal to the surface at the interface could be
into the surface, or it could be out of the surface.
The terms "blowing" and "suction" are frequently used
to denote the direction of flow at the interface, while
the word "transpiration" generally is taken to embrace
both cases.   Suction is sometimes used as a scheme for
aerodynamic boundary layer control because it is possible to inhibit or prevent boundary-layer separation
by suction.

These various types of applications suggest why
chemical, mechanical and aeronautical engineers have
all made significant contributions to the theory, and
the terminology to a certain extent reflects these
various origins of interest.

In the class of problem considered here it is
assumed that the surface is aerodynamically smooth,
and that the holes or pores in the surface through
which the transpiration fluid flows are sufficiently
small relative to the boundary layer thickness for the
velocity normal to the surface to be treated as uniform
over every small region.   (If the holes are large,
with large spacing, the boundary layer structure is
altered.   Although the resulting boundary layer may

have some of the characteristics of the transpired
boundary layer, that problem will not be considered
here.)

The transpired boundary layer may be laminar or
turbulent.   The laminar boundary layer with transpiration has been extensively studied, resulting in a
large number of exact mathematical solutions for certain fundamental cases where similarity in velocity
profiles, and in temperature profiles, is obtained,
and various approximate methods have been developed to
handle cases when similarity does not exist.   More
recently finite difference solutions have become so
easy to obtain with digital computers, for any kind of
boundary conditions, that further investigation of the
laminar boundary layer does not appear very fruitful,
except for some very special cases.   The turbulent
boundary layer, on the other hand, is not nearly so
well understood, even without transpiration.   Up until
about ten years ago there were remarkably few experimental data available for the transpired turbulent
boundary layer.   Mickley *et al.* [1,2] at MIT had, during the 1950's, studied the momentum and concentration
boundary layers;   very little had been done with the
thermal boundary layer.   Certainly there were insufficient data of adequate accuracy to provide the basis
for anything approaching a complete turbulent boundary
layer theory.   In the 1960's experimental activity
went forward with increasing intensity in various parts
of the world, and in 1965 the authors joined in this
effort.

At the present time the behaviour of the turbulent
transpired momentum boundary layer for an essentially
incompressible fluid is fairly well understood for a

sufficient range of boundary conditions to make it
worthwhile to attempt to summarize the available data,
and the state-of-the-art in boundary-layer analytic
prediction.   A similar statement can be made about the
thermal boundary layer, but only for fluids with Prandtl
number near unity.   This is not to say that predictions
adequate for most engineering design cannot be made for
other fluids, but experimental confirmatory data are
lacking.   Data on the turbulent transpired concentration boundary layer are considerably scarcer.   The
concentration boundary layer problem is complicated by
the fact that strong transpiration is usually accompanied by large fluid property gradients caused by
large concentration gradients, and this adds more variables to the experiments.   Of course if concentrations,
and thus concentration gradients, are small, and if the
Lewis number is near unity, the behaviour of the concentration boundary layer should be identical with that
of the thermal boundary layer.

In order to restrict this paper to a range of variables and boundary conditions for which there is now a
considerable body of data, the scope of the paper will
be restricted to the following:

   (a) Momentum and thermal boundary layers only
   (b) Constant fluid properties, so that the momen-
        tum and energy equations are effectively de-
        coupled
   (c) Low velocity flows, with a similar result as
        in (b)
   (d) Two-dimensional boundary layers
   (e) Steady flow
   (f) Aerodynamically smooth surface
   (g) Injection or suction velocity uniform over
        each small area of surface (though possibly
        varying in the streamwise direction on a
        larger scale)

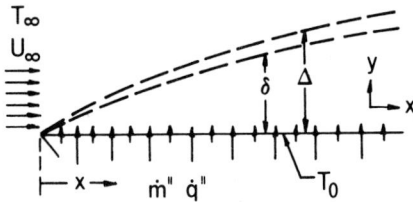

Fig. 1   The physical situation.

Within this scope, the objective of the article
will be to summarize the present status of our experimental knowledge of this family of boundary-layer
flows, and to demonstrate how these flows can be predicted by a reasonably simple mathematical model and
a finite-difference calculation procedure.   Because
the authors themselves have contributed much of the
available data to the literature, the main line of experimental data to be presented will be the authors'
own.   However, wherever practicable, comparison will
be made with the data of others, because there are indeed some discrepancies and, consequently, some differences of opinion.   The mathematical model to be proposed is only one of many eddy-viscosity/eddy-conductivit
models discussed in the literature, all of which are
actually very similar.   However, it is one that has
been used with considerable success by the authors, as
will be demonstrated.   It is not a purpose of this
article to draw critical comparison between various
prediction schemes.   It is our hope however that others
might find the experimental data summarized here to be
useful for such an evaluation.

The general problem considered is illustrated by
reference to Fig. 1.   A fluid flows at a steady rate

along a flat surface which is porous, and through
which fluid with the same composition can be forced
into the boundary layer or withdrawn from the boundary
layer (blowing or suction).   The Reynolds number is
sufficiently high so that the boundary layer is turbulent.   It is presumed that the surface is aerodynamically smooth and that the velocity normal to the surface, whether positive or negative, is uniform over an
area large relative to the sublayer thickness.   It is
presumed that the solid surface is a heat conductor,
and that heat can be conducted to or from the surface.
It is further presumed that the surface construction
is such that the transpired fluid is in thermal equilibrium with the solid surface at the interface.   In
general, our long-range objective would be to consider
the case where the mass transfer rate $\dot{m}''$ (*i.e.*, the
transpiration rate) is any arbitrary function of distance x along the surface and where the convection
heat transfer $\dot{q}''$ is likewise any arbitrary function
of distance along the surface (or the surface temperature T_0 is any arbitrary function of x).   We would
like, ultimately, to consider the case where the free-
stream velocity U_∞ may vary in any arbitrary manner
with x, but we will restrict free-stream temperature
T_∞ to a constant.   The boundary layer is two-
dimensional, with the co-ordinate y used to measure
the distance normal to the surface (*i.e.*, all properties
are uniform with respect to the z-direction).   Under
these conditions we are interested in the development
and properties of a momentum boundary layer, characterized by a thickness δ_2, and a thermal boundary layer
characterized by a thickness Δ_2.

It is apparent that solution of the general prob-

lem described above is going to require a theory incor-
porating some broadly applicable hypotheses about the
turbulent transport mechanisms. The number of independent variables, and the infinite possibilities for
varying boundary conditions, make it impractical to
consider the totally experimental approach wherein experimental data are generalized by dimensional analysis
and then applied directly to particular problems.
This raises the question as to what kinds of experiments
should be carried out to provide the experimental basis
for a general theory.   What are the more fundamental
cases that should be tested to provide firm bench-marks
and from which the various constants and functions
necessary to a more general theory can be derived?
What are some critical experiments that should be carried out to provide a severe test of a general theory?

        Experience with the laminar boundary layer, which
can of course be completely handled analytically, has
pointed the way to certain fundamental cases that also
arise with turbulent boundary layers.   Certain parameters can be maintained constant, which makes it easier
to derive the critical constants and to determine their
functional dependence.   In laminar boundary layer
theory the concept of velocity-profile similarity leads
to a very considerable mathematical simplification, and
to a whole family of simple solutions for some particular cases of transpiration and free-stream velocity
variation.   Clauser [3] demonstrated that for a turbulent boundary layer without transpiration a family of
"equilibrium" boundary layers exist which have partial
velocity profile similarity, and Bradshaw [4] demonstrated that essentially the same free-stream velocity
variation that yields the laminar boundary-layer "simi-

larity" solutions also leads to "equilibrium" boundary
layers in the turbulent case.   Anderson [5] has shown
that essentially the same situation exists for transpired turbulent boundary layers.   Thus the family of
"equilibrium" transpired turbulent boundary layers appears to provide a fundamental set, and will be used
in Sec. 3 for the main presentation of experimental
data.   The constants and functions for use in a more
general theory will be derived from these "equilibrium"
experiments.   Then, to provide some severe tests of
general theories, a small amount of data for "non-
equilibrium" cases will be presented in Sec. 4.

    In Sec. 2 below, the concept of the "equilibrium"
boundary layer will be discussed more precisely.   The
available experimental data for "equilibrium" transpired boundary layers will then be presented in three
groups: firstly the case of a constant free-stream
velocity, then the case of an accelerating free-stream
velocity (favourable pressure gradient), and finally
the case of decelerating free-stream velocity (adverse
pressure gradient).

    After presentation of some data on "non-equilibrium"
flows, mathematical models for both the momentum and
energy equations will be discussed, and the appropriate
constants and functions derived from the "equilibrium"
experiments will be presented.   Finally some examples
of predictions based on this particular theory will be
presented.

    As a final word of introduction, it should be
pointed out that although the primary subject of this
article is the turbulent boundary layer with transpiration
the case of the impermeable wall, *i.e.*, non-transpiration,
is a valid member of the family of flows considered.

The data presented for this case, which of course has
been extensively studied by many workers over the
years, was obtained in the authors' laboratory. However, in all essentials it is virtually identical with
that reported by most other workers, and thus can probably be considered as definitive.

2.   EQUILIBRIUM BOUNDARY LAYERS

The motivation which has guided the choice of
parameters for the so-called equilibrium turbulent
boundary layers has its roots in the similarity variables of laminar boundary layer theory.   In laminar
boundary layers, fixing the value of an appropriate
ratio of boundary conditions allows the reduction of
the partial differential equation to an ordinary one
for some flow conditions, and permits a relatively
simple mathematical solution.   The resulting velocity
and temperature profiles are exactly self-similar, and
there is no uncertainty as to the efficacy of the
parameters chosen: the results speak for themselves.
No such drastic benefit is realized in turbulent
studies.   Combinations of boundary conditions can be
proposed as being likely to lead to self-similar behaviour of the boundary layer, but the profiles must be
found experimentally and they are not usually exactly
self-similar, only approximately so.   Turbulent boundary layers driven by these carefully chosen combinations
of boundary conditions are known as "equilibrium" or
"asymptotic" boundary layers: not truly similar, but
closely so.
The best-known class of flows leading to laminar
similarity solutions is the Falkner-Skan family, which

results when the free-stream velocity, U_∞, varies as x^m (m positive or negative) and v_o varies in a related manner, *i.e.*,

$$U_\infty \propto x^m \tag{1}$$

and

$$v_o \propto U_\infty(c_f/2) \tag{2}$$

It is particularly important to note that similarity is not obtained in general when v_o is a constant, independent of x. The special case of $v_o = 0.0$ and U_∞ a constant does yield similarity in velocity profiles, but, in general, a constant value of v_o is an "arbitrary" variation of blowing as far as similarity is concerned. Note further that similarity is achieved only if $v_o/(U_\infty c_f/2)$ is a constant with respect to x. This dimensionless group is usually called the "blowing parameter", and will be termed B_m. Thus,

$$B_m \equiv \frac{\rho_o v_o}{\rho_\infty U_\infty(c_f/2)} = \frac{\dot{m}''/G_\infty}{c_f/2} \tag{3}$$

The physical significance of holding B_m constant can be appreciated if it is observed that (3) can be rearranged as:

$$B_m = \frac{(\rho_o v_o)U_\infty}{\tau_o} \tag{4}$$

In this form it can be seen that B_m is the ratio of the transpired momentum deficit to the surface shear force. When these are kept in a fixed ratio along a surface, then the laminar boundary layer develops in such a way as to produce similar velocity profiles.

It seems likely that this ratio would also be important
in turbulent boundary layers.

The energy equation can also be cast in such form
as to reveal its similarity variables.   Similar temperature profiles result when, in a laminar boundary
layer having hydrodynamic similarity, the wall and
free-stream temperatures are constant and the "heat-
transfer blowing parameter" B_m is held fixed.   This
parameter reflects the ratio of the transpired energy
deficit to the surface heat transfer and is defined
by:

$$B_h \equiv \frac{\rho_o v_o}{\rho_\infty U_\infty St} = \frac{\dot{m}''/G_\infty}{St} = \frac{\dot{m}''c(T_o - T_\infty)}{\dot{q}_o''} \qquad (5)$$

The blowing parameter B_m and the heat-transfer
blowing parameter B_h both arise in the reduction of
the partial differential equations of the laminar
boundary layer to the ordinary differential equation of
the similarity situation.   Both, however, are also
visible in the integral equations of the boundary layer:
a form which applies also to turbulent boundary layers.

The requirement imposed by Eq. (1) can be shown
to result in a pressure gradient parameter, β, which
remains constant when m is constant, where:

$$\beta \equiv \frac{\delta_1}{\tau_o}\left(\frac{dp}{dx}\right) \qquad (6)$$

β may be interpreted as the ratio of the axial pressure
force acting on the boundary layer to the shear force
at the wall.   Thus B_m and β should have similar
influences upon the development of the momentum boundary
layer, and indeed if one examines the following form of
the momentum integral equation of the boundary layer,
this is seen to be the case:

$$\frac{d(U_\infty^2 \delta_2)}{dx} = \left(\frac{\tau_o}{\rho_\infty}\right) (1 + B_m + \beta) \qquad (7)$$

Equation (7) expresses the rate of growth of the momentum deficit of the boundary layer. If B_m and β are held constant along a surface, it is not surprising that the boundary layer maintains a similarity of structure as it develops.

The energy integral equation can be manipulated in such a way as to show the importance of B_h:

$$\frac{d[c\Delta_2 U_\infty (T_o - T_\infty)]}{dx} = \frac{\dot{q}_o''}{\rho_\infty} (1 + B_h) \qquad (8)$$

Equation (8) expresses the rate of growth of the axially flowing energy flux in the thermal boundary layer, and B_h is seen to have the same influence on the thermal boundary layer as B_m has on the momentum boundary layer. Note, however, that the pressure gradient parameter β has no direct effect on the thermal boundary layer.

Let us now turn to the *turbulent* boundary layer. It would be convenient to be able to define some kind of similarity that would lead to a classifiable group of flows. The problem is not quite so straightforward as for laminar boundary layers, because with turbulent boundary layers there are two distinct regions to consider: the inner and the outer regions behave differently. It is possible to have *inner region similarity* independent of the outer part of the boundary layer (although the latter may extend over most of the boundary layer). The existence of a local "law-of-the-wall" which seems to hold under many conditions regardless of upstream history is witness to this fact. Put another

way, we have been discussing some conditions of ratios
of forces acting on the boundary layer that lead to
similar structure in laminar boundary layers.   With
the turbulent boundary layer it is possible for the
inner region, near the wall, to be in equilibrium while
the outer region continues to develop.   This inner-
region equilibrium appears to be associated with an
equality between the rate of production of turbulent
energy and its rate of dissipation.

Clauser [3] proposed that boundary layers having
outer region similarity be called *equilibrium boundary
layers*, and that an equilibrium boundary layer be one
for which the outer region velocity profile, plotted
in *velocity-defect co-ordinates* was universal.   This
condition can be expressed by:

$$\frac{U-U_\infty}{U_\tau} = f\left(\frac{y}{\delta_3}\right) \quad \text{only} \tag{9}$$

where $\qquad \delta_3 = \int_0^\infty \frac{(U-U_\infty)}{U_\tau} \, dy \tag{10}$

Clauser also proposed a *shape factor*, G, that
would be a constant, independent of x, under these
conditions:

$$G = \frac{1}{\delta_3} \int_0^\infty \left(\frac{U-U_\infty}{U_\tau}\right)^2 \, dy \tag{11}$$

Experimentally, it has been found that if β is
held constant, G remains constant.   Similarly, it
has been found that holding B_m constant also yields
constant G profiles.   More recently it has been
shown (Andersen [5]) that if $(B_m + \beta)$ is maintained
constant, G will be constant.   Thus it appears that
the same relationship among the forces acting on the

boundary layer that yield similarity solutions for
laminar boundary layers also yields equilibrium boundary layers, in the Clauser sense, for turbulent boundary layers.

It is not surprising that the experimental condition leading to the constant β (and thus constant
G) boundary layers is, again,

$$U_\infty \propto x^m \qquad (12)$$

If there is transpiration in addition, B_m must be
constant in order to yield constant G, and this requires that

$$v_0 \propto U_\infty(c_f/2) \qquad (13)$$

which may readily be shown to reduce, to a good approximation, to

$$v_0 \propto x^{m_F} \qquad (14)$$

where $m_F \simeq m - 0.2$ if $c_f/2 \propto Re_x^{-.2}$

Figure 2 shows an example of a series of velocity-
defect profiles for a blown adverse pressure gradient
equilibrium boundary layer for which B_m, G, β, m,
and m_F are all constant. Figure 3 shows the addi-
tive character of B_m and β for 18 equilibrium
boundary conditions.

A rather special case of equilibrium turbulent
boundary layers occurs in accelerating flows where the
acceleration parameter, K, is maintained constant.

$$K \equiv \frac{\nu}{U_\infty^2} \frac{dU_\infty}{dx} \qquad (15)$$

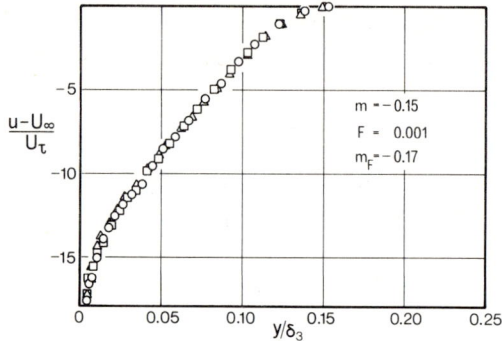

Fig. 2   Defect profiles for an equilibrium boundary
layer.

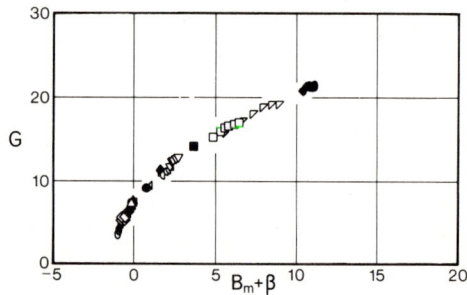

Fig. 3   The Clauser shape factor, G, versus B_m + β.

The significance of constant K can be appreciated if the momentum integral equation of the boundary
layer is written in the following form:

$$\frac{dRe_m}{U_\infty dx/\nu} = c_f/2 + v_o/U_\infty - K(1 + H)Re_m \qquad (16)$$

If K and v_o/U_∞ are maintained constant, independent of x, and if K is finite and *positive*, the
flow must inevitably approach a state of equilibrium
for which Re_m is constant.   This is often spoken of

as an asymptotic accelerating flow; a special case of
an equilibrium boundary layer in which there is not
only outer region similarity (constant G), but also
inner region similarity.   The velocity profiles are
similar all the way to the wall, with the result that
not only is Re_m constant, but so also are $c_f/2$ and
the shape factor H.   Thus, constant K and constant
v_O/U_∞ together yield a family of similarity solutions
for laminar boundary layers and a family of asymptotic-
accelerating layers for the turbulent case.   An
interesting feature of Eq. (16) is the fact that for
each positive value of K and each value of F there
exists a definite value of Re_m: as K increases,
Re_m decreases.   Experiments indicate that it is impossible to maintain a turbulent boundary layer if
Re_m is below about 300.   The corresponding value
for K is about 3×10^{-6}.   In other words, if K
is of the order of 3×10^{-6}, or greater, the turbulent boundary layer will tend to revert to a laminar
boundary layer.   Evidence of this trend will be demonstrated in some of the experimental data to be presented.

    If K is negative (*i.e.*, a decelerating flow),
no such asymptotic equilibrium can exist (except as
discussed below).   Note also that for a given value
of K the rate of transpiration, whether positive or
negative, will have a substantial influence on the
asymptotic value of Re_m.

    Another related type of asymptotic flow can be
recognized in Eq. (16).   If v_O/U_∞ is negative, Re_m
will approach a constant when K is zero or negative,
so long as the v_O/U_∞ term is larger in the absolute
sense than the last term.   This type of boundary layer

is frequently referred to as the "asymptotic suction
layer", and may be either laminar or turbulent, depending upon the magnitude of Re_m at the asymptotic condition.   Note that for $K = 0.0$, $c_f/2$ approaches an
asymptote, $-v_o/U_\infty$.   Physically, the surface shear
force is then precisely equal to the loss of momentum
of the fluid that is brought from the free-stream to
zero velocity at the surface.

The energy integral equation of the boundary layer
can be put in a form similar to that of Eq. (16), for
the case of constant wall and free-stream temperatures.

$$\frac{dRe_h}{U_\infty dx/\nu} \;=\; St + v_o/U_\infty \tag{17}$$

The important difference is that there is no term corresponding to the acceleration term (or pressure gradient term).   Thus if K is a positive constant, the
momentum boundary layer will come to equilibrium, with
Re_m constant, but the thermal boundary layer will con-
tinue to grow.   In fact if K is maintained constant
for sufficient distance the thermal boundary layer will
grow outside of the momentum boundary layer since Re_h
increases indefinitely.

As Eq. (17) suggests, an asymptotic thermal boundary layer exists in the case of negative v_o/U_∞:   an
"asymptotic-suction-layer" similar to the momentum
boundary layer case.

The general question of equilibrium turbulent
thermal boundary layers, *i.e.*, thermal boundary layers
having outer region temperature profile similarity,
has not, so far as the authors are aware, been systematically explored, although Blackwell [6] has studied
adverse pressure gradient equilibrium boundary layers

with constant surface temperature.   If surface temperature is allowed to vary, additional possibilities
for similarity arise.   In order to restrict this
paper to a few classifiable types of boundary layers,
consideration will be restricted to thermal boundary
layers on constant temperature surfaces, with a constant temperature free-stream.

There is no question that the turbulent thermal
boundary layer which forms on a constant temperature
surface when the free-stream velocity is constant does
have outer-region similarity.   There is also a fixed
ratio (for a given Prandtl number) of enthalpy thickness Reynolds number to momentum thickness Reynolds
number.   Even when the boundary layer starts with
widely differing enthalpy and momentum thickness
Reynolds numbers, this ratio of sizes will be approached
at points downstream.

When there is blowing or suction with constant
free-stream velocity, and the blowing parameter B_m
is maintained constant, it is observed experimentally
that B_h will also approach a constant.   The thermal
boundary layer will approach an equilibrium state with
a fixed ratio of enthalpy to momentum Reynolds numbers,
regardless of the starting conditions.   Blackwell [6]
observed the same situation for an adverse pressure
gradient flow for which β was held constant.   Examination of the integral equations, (7) and (8), might
suggest that analogous behaviour for both boundary
layers would be unlikely when there is a pressure
gradient, because β appears in only one of the
equations.   However, the adverse pressure gradient is
accompanied by a decrease in the surface shear stress
(and in $c_f/2$),   but no significant change in heat flux

(and   St),   with the result that the rate of growth of
both boundary layers becomes the same.   A similar
conclusion results from examination of Equations (16)
and (17).   For an equilibrium, or constant β,   boundary layer, the final term in Eq. (16) tends to vary
with   x   at the same rate as does $c_f/2$,   and   St   in
Eq. (17) also varies at the same rate.

The equilibrium thermal boundary layer can be defined in an analogous manner to the equilibrium momentum boundary layer, *i.e.*, a thermal boundary layer
having outer-region similarity of the temperature profile.   Following the form of Equations (9) and (10),
temperature defect coordinates are defined,

$$ t_d^+ = \frac{(T_\infty - T)}{(T_\infty - T_o)} \frac{\sqrt{c_f/2}}{St} = f\left(\frac{y}{\Delta_3}\right) \tag{18} $$

where

$$ \Delta_3 = \frac{\sqrt{c_f/2}}{St} \int_0^\infty \frac{(T_\infty - T)}{(T_\infty - T_o)} \, dy \tag{19} $$

Figure 4 shows three temperature profiles plotted
in these coordinates for an adverse pressure gradient,
the momentum boundary layer having velocity-defect
similarity, *i.e.*, an equilibrium momentum boundary
layer.   Note that the temperature profiles are universal everywhere outside of the sublayer.   Thus an equilibrium momentum boundary layer with   G, β,   and   m
all constant will also yield an equilibrium thermal
boundary layer when the surface temperature is constant.
The same is true if there is transpiration with B_m
constant: B_h will become constant and similar temperature profiles result.

This behaviour might also be anticipated from ex-

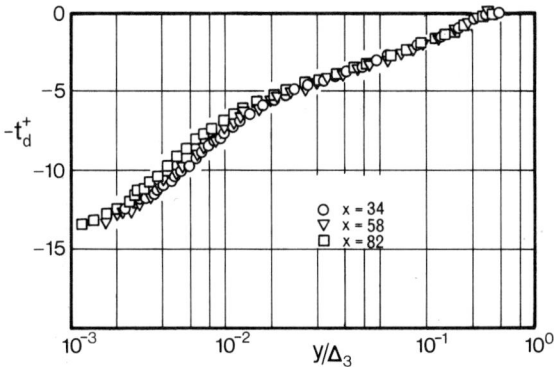

Fig. 4   Temperature defect profiles for a decelerating flow, m = -0.2.

perience with the similarity solutions for the laminar
boundary layer.   Recall that the condition of Eq. (1)
leads to laminar similarity solutions for the thermal
boundary layer, so it is not surprising that this condition leads also to equilibrium turbulent thermal
boundary layers.

In summary, equilibrium turbulent momentum boundary layers, having outer-region velocity-profile similarity, can be achieved by holding constant the pressure
gradient parameter, β, and the blowing parameter,
B_m.   Experimentally, these conditions can be established by adjusting free-stream velocity, U_∞, and
transpiration velocity, v_o, according to the condition of Equations (1) and (2).   A special case of
momentum-boundary-layer equilibrium is approached for
accelerating flows when the acceleration parameter,
K, is maintained finite and positive.   For this
special case the velocity profiles possess both inner
and outer similarity.

Equilibrium thermal boundary layers are obtained,
if surface and free-stream temperature are maintained
constant, under the same conditions leading to equili-

brium momentum boundary layers.   The special case of
a constant acceleration parameter,   K,   does *not* result
in an equilibrium thermal boundary.

3.   EXPERIMENTAL RESULTS

3.1   *Preliminary Remarks*

The objective of this section is to illustrate
the hydrodynamic and heat-transfer behaviour of equilibrium and near-equilibrium turbulent boundary layers
subject to blowing and suction, acceleration and deceleration of the free stream.   These are the cases
used to generate the differential correlations used in
the mathematical model to be described later.   They
show the principal physical responses of the boundary
layer and serve to illustrate the complexity of the
prediction problem.   No attempt has been made to present a complete survey of the available data.   The
reader interested in the complete data sets should ac-
quire the referenced theses, which contain full, tabu-
lar, data sets.

The Stanford experimental results all derive from
the same basic apparatus, described by Moffat [7] and
modified by subsequent authors to permit studies of
accelerating and decelerating flows.

The apparatus is an open-circuit wind tunnel whose
test section is 8 feet long, 20 inches wide and approximately 6 inches high at the inlet end.   The lower surface of the test section is the porous working plate
while the top is a control surface and can be adjusted
to vary the free-stream velocity in the streamwise
direction.   The porous plate is subdivided into 24

strips each 4.0 inches long in the flow direction and
each provided with a transpiration flowmeter, a set
of imbedded electric heaters, and temperature measuring
instrumentation. Heat transfer rates are deduced
from the measured electrical power by energy balance,
accounting for the heat losses from the porous strip
to its surroundings by conduction and radiation. All
tests were conducted with air as the transpired fluid
as well as the main stream fluid, and with small temperature differences (approximately 25°F) to keep the
effects of properties variation through the boundary
layer negligible.

The porous plates are known to be uniform within
±6% in permeability in the centre 6-inch span, where
the data are taken. The plates were custom made by
sintering bronze particles together in a polished
stainless steel mould cavity. The particles ranged
in size from 0.002 to 0.005 inches in diameter. The
fabrication technique resulted in a porous material
whose surface feels smooth to the touch and which dis-
plays a roughness of 250 micro-inches on a standard
Surfa-gauge test. The plates were installed with an
insulating spacer of 0.020 thickness between adjacent
plates, with the joints finished by hand until they
were not discernible to the touch.

Wall temperatures were held uniform to within
±0.5°F in cases reported as "constant wall temperature"
to reduce conduction transfer between plates. The
main stream velocity was shown to be uniform in the
spanwise direction within ±0.4% with the free stream
turbulence intensity being about 0.5%. Spanwise tests
of the boundary-layer development showed the momentum
thickness to be uniform within ±2% across the measuring

portion of the tunnel.   Three-dimensional effects on
the heat transfer measurements were investigated by
comparing integrals of the enthalpy flux in the boundary layer with heat-transfer integrals along the plate.
Simpson [8] concluded that the 3-D effect was not
larger than 3%.   It follows from the similarity between the energy and momentum equations that three-
dimensional effects on skin friction data should be of
the same order of magnitude.

Statements regarding the "relative uncertainty"
of Stanton number have little meaning since blowing
tends to force the Stanton number to zero.   Absolute
uncertainties have more meaning, and will be referred
to here.   The stochastic component of uncertainty in
reported values of Stanton number is about ±0.0001
Stanton number unit.   This value was estimated by
applying constant probability propagation of uncertainty to the data reduction program using standard
uncertainty values for the input measurements.   Values
of the skin friction coefficient are more difficult to
measure than heat transfer and tend to be less certain.
In addition, the skin friction data tend to be sparse
(for example, the Stanford results present only 4 or
5 values of skin friction per situation whereas 24
values of Stanton number are presented), and therefore
admit of larger differences in opinion in interpretation.   Thus, although the stochastic component of
uncertainty in each measurement may be ±5%, there may
be differences in interpretation of the same data by
different investigators due to differences in curve-
fitting coordinates *etc*. which are as large as 10% or
15% (see Squire, [9]).

3.2   The Experimental Boundary Conditions

The overall program from which the present examples of data were taken covered accelerating and decelerating flows as well as variations in wall temperature and blowing.   This range precluded the use of x-Reynolds number as a useful correlating parameter and soon led to the use of the local boundary thickness Reynolds number (either momentum- or enthalpy-thickness) for presentation of the results.   It was shown, by the study of Whitten [10], that for uniform free-stream velocity, the boundary layer would adjust to even a step change in blowing within 2-5 boundary-layer thicknesses.   It can be seen by comparing his data with those of Moffat [7], that uniform blowing (constant   F flows) produced the same value of Stanton number for a given enthalpy thickness Reynolds number as did constant   B   flows.   With the validity of "local equilibrium" established to at least this extent, it was possible to conduct the remaining experiments with the experimentally simpler boundary condition of constant F   rather than the true equilibrium situation of constant   B.   It is believed that any effects of this simplification will be found only in the velocity and temperature profiles, and then only in the outermost parts (*i.e.*, the wake region), since the inner region responds so rapidly to the wall conditions.

Accelerations were characterized by a constant value of the parameter   K.   As remarked in Sec. 2, such flows are of fundamental value;   they also have the merit of being easy to establish since they occur naturally in the flow through a channel formed between two converging planes (except for second order effects

due to the growth of the displacement thicknesses).
It was first inferred from Eq. (16), and then confirmed
experimentally, that the asymptotic state would result
in a constant value of $c_f/2$.   This has the effect
that constant F is also constant B hence, although
the experiment was set up as constant F, it resulted
in a completely equilibrium situation with both K
and B virtually constant in the accelerating region.

The decelerating flows were set to near constant
β pressure gradients by fixing $U_\infty = U_1 x^m$ with m
negative and x determined with respect to the virtual
origin.   Blowing was set by specifying constant F,
relying upon the trend towards local equilibrium.   The
results showed no discernible effect of deceleration
upon the relationship between Stanton number and enthalpy thickness Reynolds number: the same correlations which worked for the flat plate case works for
constant β decelerations.   Hence, it seems safe to
assume that the St∿Re_n relationship determined by an
experiment at constant F (quasiequilibrium) is the
same as that which would have been found in a (strictly
equilibrium) constant-B experiment.

Constant wall temperature was used throughout the
base-line tests reported here, with an average temperature difference of about 25°F between the free stream
and the wall.   No correction was made for the effect
of variable fluid properties except to the initial
"qualification runs" which validated the apparatus in
the flat-plate, no-blowing test case.   Some tests were
made to document the response of the boundary layer to
steps and ramps in temperature and to arbitrary distributions of blowing: these serve as check runs, against
which one can test models of boundary layer behaviour

based upon the differential and integral correlations derived from the equilibrium experiments.

3.3 *Correlations of Results*

The principal correlations deduced from the present data set are those required as inputs to the differential momentum and enthalpy equations, such as the mixing-length and the turbulent-Prandtl number distribution across the boundary layer. There is, however, also a need for integral correlations such as those relating the Stanton number to the enthalpy thickness Reynolds number, F, and K or β. The range of validity of such integral correlations is necessarily less than that of the "differential" correlations, since the latter have the differential equation to help cope with the changing boundary conditions. Within their limited range and accuracy, however, such correlations are extremely useful, particularly when kept to simple functional forms.

The results of the present data set are, therefore, presented in both ways: integral and differential correlations. Discussion of the differential correlations is deferred until Sec. 4. The integral correlations are discussed with the data since they help to clarify the organization of the results.

3.4 *Heat Transfer Results for* U_∞ = *Constant*

The principal effects of transpiration on heat transfer through a constant-velocity boundary layer are shown in Figures 5-7.

Figure 5 shows Stanton number as a function of x-Reynolds number for uniform blowing and suction. It is apparent that blowing $(F_\infty > 0)$ and suction $(F_\infty < 0)$

St

0·1×10³

10⁵   2   4 6 10⁶   2   4 $\dfrac{U_\infty x}{\nu}$

F

-0·0076
-0·0046
-0·0024
-0·0012
0·000
+0·001
+0·0019
+0·0038
+0·0096
+0·0078

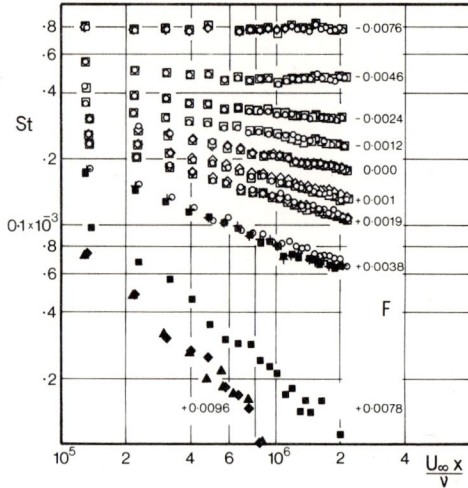

Fig. 5   The variation of Stanton number with x-
Reynolds number at constant F.

both have large effects: blowing tends to drive the
Stanton number towards zero as the blowing fraction
increases whereas suction tends to force the Stanton
number to an asymptotic value numerically equal to
(-F).   This behaviour is illustrated in Fig. 6, which
shows St as a function of F parametric in x-Reynolds
number.   For this range of Reynolds numbers, the Stanton number has, in all cases been reduced virtually to
zero when F is as large as 0.01.

Within the range of values covered by the present
data, the effect of blowing can also be described by
a correlation in Stanton-number coordinates.   The data
show that the quantity St/St_o, denoting the ratio of
the Stanton number with blowing to that without blowing
(at the same Reynolds number), is a unique function of
the blowing parameter B.   The comparison can be made
at the same x-location (*i.e.*, the same Re_x) as given

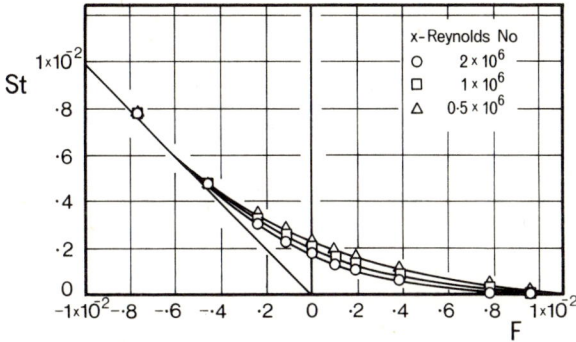

Fig. 6   The variation of Stanton number with F
            at constant x-Reynolds number.

by Equations (20) and (21):

$$\frac{St}{St_o}\bigg|_{Re_x} = \frac{ln(1 + B_h)}{B_h} \qquad (20)$$

or, equivalently,

$$\frac{St}{St_o}\bigg|_{Re_x} = \frac{b}{e^b - 1} \qquad (21)$$

where

$$B_h = \frac{\dot{m}'}{G_\infty St} \qquad (22)$$

and

$$b = \frac{\dot{m}''}{G_\infty St_o} \qquad (23)$$

Equations (20) and (21) were presented during the
early 1950's by several workers in the field as "stagnant film theory" or "Couette flow" models.   The
agreement between the data and Eq. (20) is shown in
Fig. 7.   In part, the good agreement results from the
implicit nature of Eq. (20);   diminishing St   reduces
both St/St_o and $ln(1 + B_h)/B_h$,   hence promoting
"good agreement".

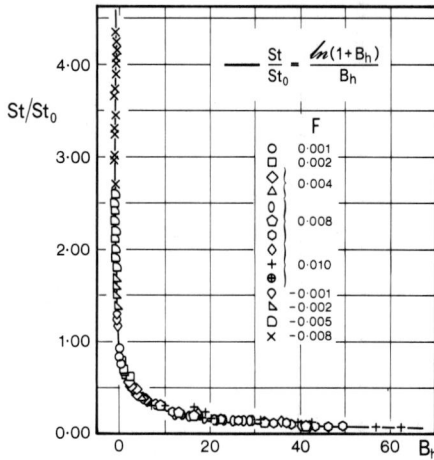

Fig. 7   The ratio St/St_O versus the heat transfer blowing parameter, B_h.

As mentioned earlier, the objective of the overall research programme suggested an early search for local descriptors of boundary-layer behaviour. In particular, emphasis was focused upon the relationship between Stanton number and enthalpy thickness Reynolds number. It was natural, then, to seek a way of predicting the effect of blowing in local coordinates: *i.e.*, St/St_o at constant enthalpy thickness Reynolds number.

The data of Fig. 5 have been re-cast to show Stanton number versus enthalpy-thickness Reynolds number and plotted as Fig. 8. Values of enthalpy thickness for this plot were calculated from the measured Stanton number data and the blowing fraction by integrating the two-dimensional energy integral equation (17) along the plate but values derived by this method agree well (within 6%) with values deduced by traversing the boundary layers. An empirical formula could be deduced from

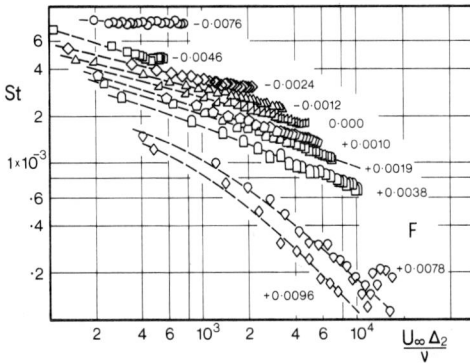

Fig. 8   The variation of Stanton number with en-
           thalpy thickness Reynolds number at con-
           stant F.

this figure but a better guide is at hand.   Whitten
[10] showed that Eq. (20) could be combined with the
two-dimensional energy integral equation and with an
equation describing the variation of St_o with Re_x
to yield the following form:

$$\left.\frac{St}{St_o}\right|_{Re_h} = \left[\frac{ln(1 + B_h)}{B_h}\right]^{1.25} (1 + B_h)^{0.25} \quad (24)$$

    Equation (24) was developed from Eq. (20), hence
all of the data for U_∞ = constant fit Eq. (24).   Implicit in Eq. (24) is the notion of local equilibrium:
it is presumed that knowledge of Re_h and B will
fix St/St_o, regardless of the upstream history.   The
validity of this hypothesis was tested by experiments
in the vicinity of a step change in blowing.   An
example of such data is shown in Fig. 9.   The boundary
layer is seen to respond very rapidly to the step, with
Stanton number dropping almost all the way from the un-

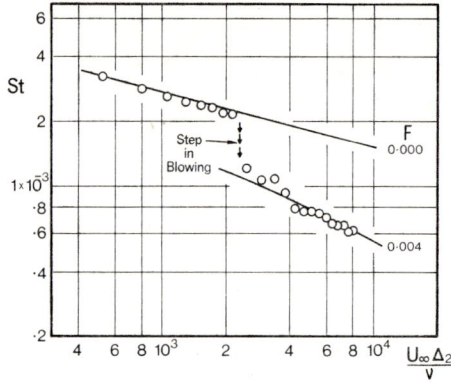

Fig. 9   The response of Stanton number to a step
           increase in blowing.

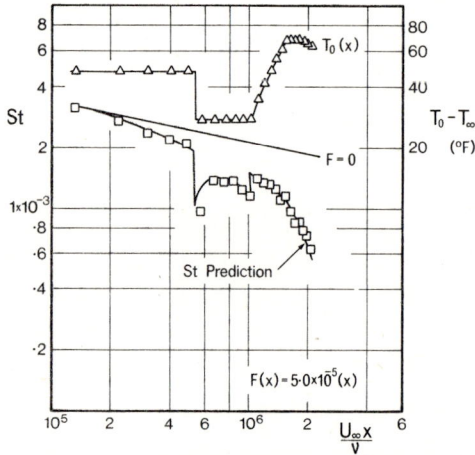

Fig. 10   A demonstration case: linearly varying
            blowing with arbitrary wall temperature.

blown value to the uniformly blown value within the
width of one porous section (4 inches in the flow
direction).   The boundary-layer thickness at the
point of the step (99% velocity thickness) was approximately one inch hence local equilibrium is re-estab-

lished in about four boundary-layer thicknesses. This
rapid approach to "local equilibrium" supports the use
of Eq. (24) for cases of variable as well as uniform
blowing; for many problems it is found that sufficient
accuracy is obtained by such local predictions of boundary-layer behaviour. An example of a more complex
case is shown in Fig. 10, in which a linearly increasing blowing rate, $F = 5 \times 10^{-5}(x)$, was combined with
a sharply variable wall temperature. Also shown is
the predicted Stanton number variation due to Whitten
[10] which was obtained by using a local-equilibrium
formula in conjunction with a superposition method;
agreement with experiment is virtually complete.

3.5 *Skin Friction Results for* U_∞ = *Constant*

The principal effects of transpiration on the
skin-friction coefficient of a turbulent boundary layer
are illustrated in Figures 11 and 12. Figure 11 shows
the early results of Simpson [8], derived from pitot
probe surveys of the boundary layer. Figure 12, in
Re_m coordinates, includes later results of Andersen
[5] for comparison. As can be seen, there are considerable differences in the results. Our present
opinion is that the values near the bottom of each
band are more representative than those near the top.
Andersen's data were taken with a rotatable, slanted,
hot wire probe measuring turbulent shear stress near
the wall and extrapolating to the wall using a partially integrated form of the boundary layer momentum
integral equation. His data are less sensitive to
interpretive differences than were Simpson's, who determined $c_f/2$ by differentiating a curve fit through
measured values of momentum thickness. As previously

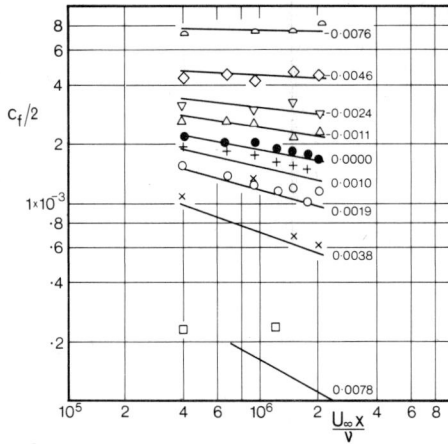

Fig. 11   The variation of friction factor, $c_f/2$, with x-Reynolds number at constant F.

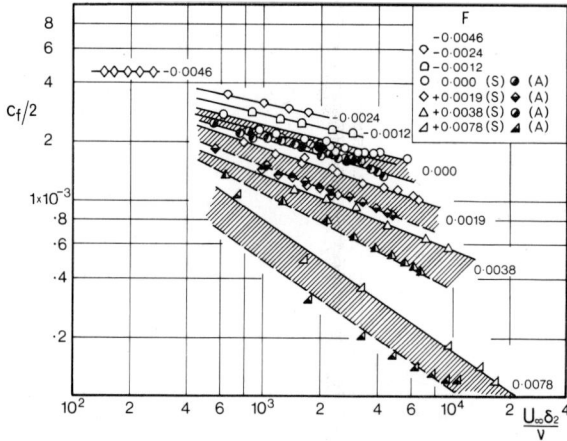

Fig. 12   The variation of friction factor, $c_f/2$, with momentum thickness Reynolds number at constant F.

mentioned, Squire [9] pointed out that Simpson's momentum thickness data could be interpreted to show different values of $c_f/2$, by 10% or more, simply by choosing different coordinates in which to curve fit the momentum thickness variation.

The variation of $c_f/2$ with blowing is essentially the same as that of Stanton number: the ratio c_f/c_{f_o} can be calculated from the blowing parameter B_m, where the subcript signifies B defined with $c_f/2$ instead of St.

$$\left.\frac{c_f}{c_{f_o}}\right|_{Re_x} = \frac{ln(1 + B_m)}{B_m} \tag{25}$$

$$\left.\frac{c_f}{c_{f_o}}\right|_{Re_m} = \left[\frac{ln(1 + B_m)}{B_m}\right]^{1.25} (1 + B_m)^{0.25} \tag{26}$$

where $\qquad B_m = \dfrac{F}{c_f/2} \tag{27}$

Judging by the rapid rate of response of St, it is to be expected that skin friction would adjust quickly to a step change in blowing; no sufficiently complete data seem to be available at present, however, for this expectation to be confirmed.

3.6 Velocity and Temperature Profiles

Velocity and temperature profiles are shown in Figures 13 and 14 for the case of constant velocity (about 40 ft/s), constant wall temperature (ΔT about 25°F), injection of air into air; the so-called "inner-layer" coordinates, u^+ ($\equiv u/U_\tau$) and y^+ ($\equiv yU_\tau/\nu$) are used. The general features of the region from $y^+ = 10$ to $y^+ = 100$ can be deduced from a Couette-

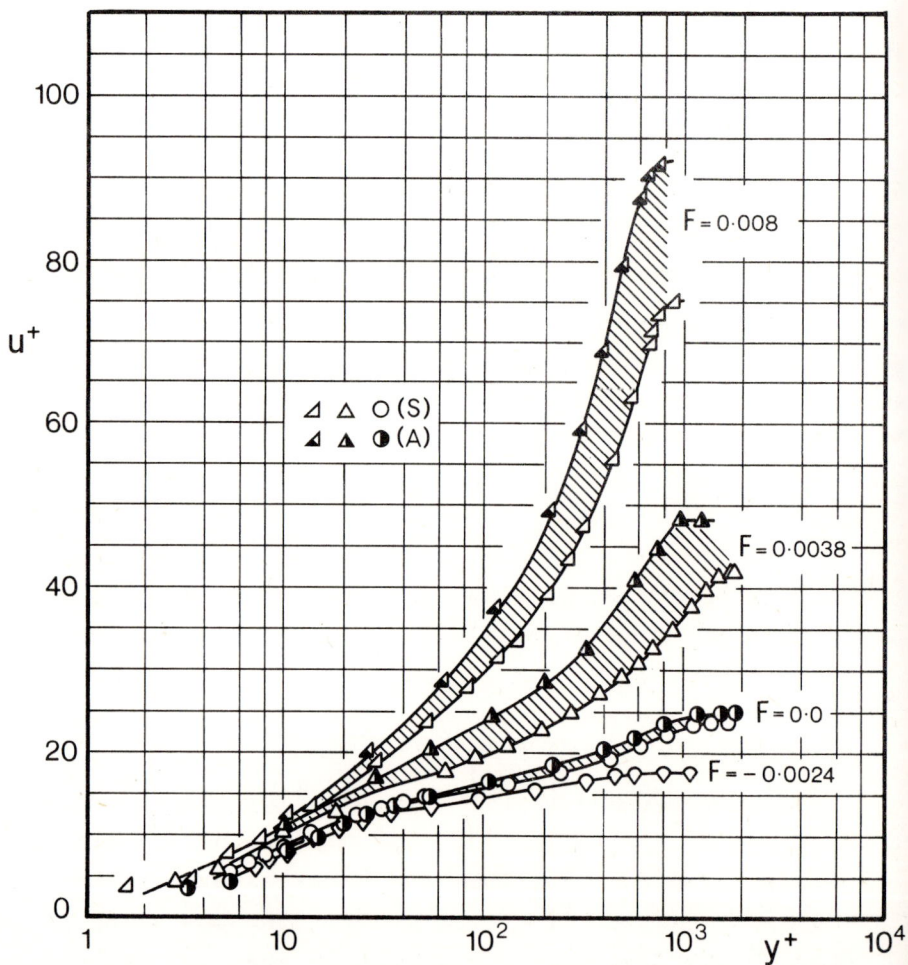

Fig. 13   Velocity profiles with blowing and suction,
           constant F and U$_\infty$.

flow analysis using a mixing-length model assuming no
effect of blowing on the mixing length distribution.
Such an analysis leads to a closed form "law-of-the-
wall" representation for the fully turbulent region.
Analyses of this sort have been presented by Black
and Sarnecki [11], Stevenson [12], Simpson [8], and
others.   They indicate that the dramatic turning up-
ward of the velocity profiles in the outer regions reflects mainly the effect of transpiration on the shear
stress distribution in the layer - not a drastic change
in the mechanism of momentum transfer.

Figure 13 shows data from two different projects
(Simpson's and Andersen's) which used different methods
of evaluating the wall shear stress (which appears in
both the u^+ and the y^+ coordinate definition).
The difference shown is, again, due to the difference
in reported values of the friction factor.   Our present opinion favours values near the high sides of the
bands shown.

Data for $y^+ < 10$ are suspect because of the
possibility of probe errors due to wall displacement
effects and shear effects.   No definitive studies
have been made concerning probe corrections in the
presence of transpiration, hence no corrections were
made to these data.   To some extent, the situation
is ameliorated by the fact that the finite-difference
program is only required to "bridge the gap" between
$y^+ = 0$ ($u^+ = 0$) and $y^+ = 10$ (u^+ known) by some
reasonable means to get into a region of comparatively
well-known behaviour.

Relatively little has been done in terms of
measuring the temperature distributions in turbulent
boundary layers with transpiration.   Figure 14 shows

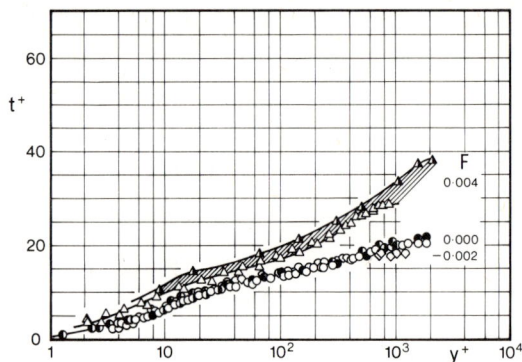

Fig. 14　　　Temperature profiles with blowing and suction, constant F and U_∞.

some results of Moffat [7], Blackwell [6], Thielbahr [13], and Kearney [14] for some cases of blowing and suction. The parameter t^+ is defined as follows:

$$t^+ \equiv \frac{T - T_o}{T_\infty - T_o} \frac{\sqrt{\frac{c_f}{2}}}{St} = \overline{T} \frac{\sqrt{\frac{c_f}{2}}}{St} \qquad (28)$$

Note that t^+ includes $c_f/2$ as well as St in its definition, hence is sensitive to the hydrodynamics as well as the heat transfer. This form follows from a Couette-flow analysis in which the terms are made dimensionless using U_τ. The fact that t^+ includes both $c_f/2$ and St means that t^+ profiles have inherently more scatter than u^+ profiles. In fact, if one examines Eq. (28) in terms of an uncertainty analysis on a simple product form, at any value of y^+, the value of t^+ includes an uncertainty component due to $\overline{T}$, a component due to $\sqrt{c_f/2}$, and one due to St. If St is uncertain within ±5%, $c_f/2$ is uncertain within 10%, and $\overline{T}$ within 2%, then

$$\frac{\delta(t^{+})}{t^{+}} = \left\{ \left(\frac{\delta\overline{T}}{\overline{T}}\right)^{2} + \frac{1}{2}\left(\frac{\delta(c_{f}/2)}{c_{f}/2}\right)^{2} + \left(\frac{\delta(St)}{St}\right)^{2} \right\}^{\frac{1}{2}} \quad (29)$$

we have

$$\frac{\delta(t^{+})}{t^{+}} = \{.0004 + .0025 + .0025\}^{\frac{1}{2}} = 0.073 \quad (30)$$

In the cases with high blowing, $\delta(St)$ and $\delta(c_{f}/2)$ remain relatively fixed (or perhaps even become larger) while the values of St and $c_{f}/2$ approach zero, yielding large percentage inaccuracies in St and $c_{f}/2$. These are propagated immediately into t^{+}.

In short, one should be more cautious in attributing significance to the details of t^{+} variations than to u^{+} variations, because of the added uncertainty involved. Since the major uncertainties involved appear in multiplicative terms, the slopes of the t^{+} - y^{+} figures are affected when shown in the conventional semi-logarithmic coordinates.

When air is both the working and the transpired fluid, the values of Prandtl number and turbulent Prandtl number are both near unity. It would be expected as a consequence, that $\overline{u}$ and $\overline{T}$ would be similarly distributed within the boundary layer in a constant velocity flow. This is indeed the case, as a review of the data of Moffat [7] or Whitten [10] will show. In view of this, it is reasonable to expect t^{+} to vary like u^{+} within the layer.

If, throughout the boundary layer the normalised velocity and temperature profiles may be assumed completely similar, *i.e.*:

$$\frac{u}{U_\infty} \equiv \bar{u} = \bar{T} \equiv \frac{T - T_0}{T_\infty - T_0} \qquad (31)$$

then, from the definitions of u^+ and t^+ it is readily deduced that

$$t^+ \simeq u^+ \left(\frac{c_f/2}{St} \right) \qquad (32)$$

Thus, for the case of constant free stream velocity, we should not find great differences in the profiles, in these coordinates.

3.7   *Flows Subject to Acceleration*

Early in the 1960's it was observed by several studies that the Stanton number was dramatically reduced by a strong acceleration.   The relationship between Stanton number and enthalpy thickness Reynolds number strongly resembled the behaviour expected of a laminar boundary layer.   As a result of this similarity the phenomenon was labelled "laminarization" and occupied a number of workers throughout the late 60's and early 70's.   There was general agreement that a suitable acceleration parameter could be taken as

$$K \equiv \frac{\nu}{U_\infty^2} \frac{dU_\infty}{dx} \qquad (33)$$

though some felt that a better form would include $c_f/2$ to some power in the denominator.   We used K, given by Eq. (33), as the acceleration parameter.   As has been shown in Sec. 2, constant-K boundary-layer flows offer a possibility for asymptotic or equilibrium boundary layers.   Such a possibility is attractive, experimentally, on three counts: (1) the experimental

realisation of such a flow is relatively easy to
accomplish (a constant-K flow can be achieved using
convergent planar walls); (2) it produces what is
possibly a simpler family of responses by the boundary
layer, with a better chance of revealing the fundamental effects; and (3) it helps to resolve the
dilemma of which possible cases involving pressure
gradient and mass transfer, out of the infinite number
of possibilities, to choose. Evidence of "local"
behaviour already mentioned, suggests that slowly varying K conditions can be treated as quasi-equilibrium
states.

3.8 Heat Transfer to an Accelerated Flow

Figures 15-18 show the effects of acceleration.
on heat transfer for values of K between 0.57×10^{-6}
and 2.55×10^{-6} with transpiration rates adjusted
to give a uniform F along the surface. The intention was to achieve and hold the asymptotic accelerated
state for as long as possible, hence it was desirable
to start the acceleration at the particular value of
momentum thickness Reynolds number corresponding to
the asymptotic value for the values of K and F
being used. The momentum integral equation, using a
simple "flat plate" correction for friction factor,
was used as a guide to choosing the starting value of
momentum thickness Reynolds number. The efficacy of
this approach can be judged from Fig. 21 which shows
that the momentum-thickness Reynolds number did, in
fact, remain substantially constant throughout the
test section in a typical run.

In every case shown in Figures 15-18 a flat plate,
turbulent boundary layer with uniform transpiration

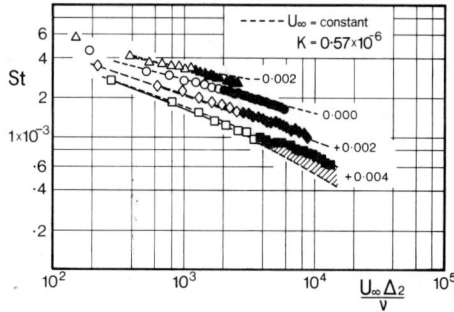

Fig. 15   The variation of Stanton number with
           enthalpy thickness Reynolds number at con-
           stant F for mild acceleration.

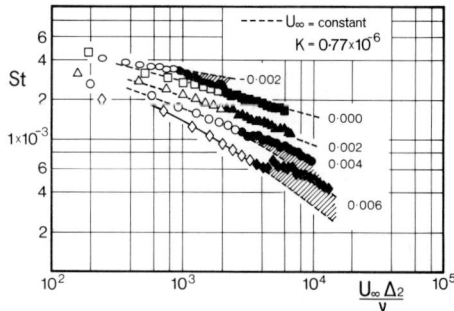

Fig. 16   The variation of Stanton number with
           enthalpy thickness Reynolds number at con-
           stant F for moderate acceleration.

was established in the test section and allowed to
grow with length until the desired momentum thickness
Reynolds number was reached.   At that location, the
top wall of the test section was adjusted to produce
in the flow the convergence required to yield the desired value of K.   Total and static pressure measurements were made at 4-in intervals along the test section to check that K was, in fact, constant through-

out the test region.

It might have been argued that unwanted effects of high velocity flow were making themselves felt in these tests besides those due to acceleration alone. Flat plate tests were conducted at velocities up to 126 ft/s to ensure that the Stanton number remained the same function of enthalpy thickness Reynolds number at the high velocity end of the test section as at the low velocity end.   At 126 ft/s the Stanton-number correlation was indistinguishable from its values at 40 ft/s though the friction factor deduced from momentum balance was higher by 5-7%.   Thus it is felt that there are no contaminating effects present: the changes in Stanton number shown in these results are those due to the acceleration level, not the velocity level.

Figures 15-18 show Stanton number versus enthalpy thickness Reynolds number for the different cases studied, compared with reference curves for flat plate behaviour.   As a general comment, for a given enthalpy-thickness Reynolds number, acceleration combined with suction reduces Stanton number, while acceleration combined with blowing increases Stanton number with respect to the transpired flat-plate correlation.   To illustrate this trend, note the progression of Stanton number behaviour for $F = -0.002$ (moderate suction) shown in the four figures.   At $K = 0.57 \times 10^{-6}$ the Stanton number slowly falls away from the flat plate result, being low by about 10% at the end of the test section.   At $K = 0.77 \times 10^{-6}$ the decline is more pronounced, with the terminal value low by almost 20%. At $K = 1.45 \times 10^{-6}$ the drop is nearly 40%, and the boundary layer seems to return only slowly to its flat

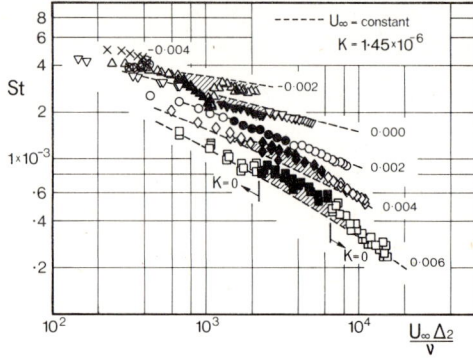

Fig. 17   The variation of Stanton number with
            enthalpy thickness Reynolds number at con-
            stant F   for strong acceleration.

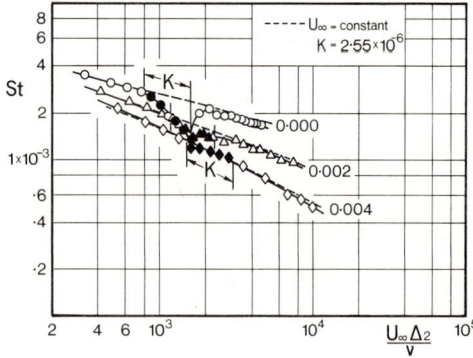

Fig. 18   The variation of Stanton number with
            enthalpy thickness Reynolds number at con-
            stant F   for very strong acceleration.

plate behaviour.   For suction, the stronger the acceleration the greater the depression of Stanton number.

     With blowing at F = +0.004, a more complex
change in behaviour is noted.   With $K = 0.57 \times 10^{-6}$
the Stanton number values rise above the flat plate

case by as much as 40% at the end of the acceleration,
and at $K = 0.77 \times 10^{-6}$ the elevation has reached
66%. For larger values of K, however, the behaviour
returns toward the flat plate correlation: at
$K = 1.45 \times 10^{-6}$ the elevation is only 35% and at
$K = 2.55 \times 10^{-6}$ the data lie once more on the flat
plate correlation. A general "trade-off" can be inferred, between the laminarizing effect of acceleration and an apparent destabilizing effect of blowing.
For positive values of F and K, the neutral values
seem to lie along a line relating K and F such
that if F exceeds $1.5 \times 10^{3}K$, the value of Stanton
number will be increased by the joint effect, and if
F is less than the neutral value, Stanton number will
be reduced.

Figure 17 also shows data for the case $F = -0.004$
(strong suction) and $K = 1.45 \times 10^{-6}$ (moderate acceleration). The trajectory of the data shows that
an asymptotic suction layer was attained for these
conditions. The first few data points show Stanton
number diminishing from 0.005 to 0.0045, in the approach region, in a typical turbulent suction layer
fashion. The acceleration begins at an enthalpy
thickness Reynolds number of 400, and Stanton number
immediately begins a sharp drop. With suction at
$F = -0.004$, the condition of thermal equilibrium at
the surface requires that the Stanton number be at
least as large as $-F$ and the decline of Stanton number is stopped at that level. With Stanton number
numerically equal to $-F$, and a constant wall temperature, the energy content of the boundary layer ceases
to change. The increasing values of U_{∞} with distance then slowly drop the value of enthalpy-thickness

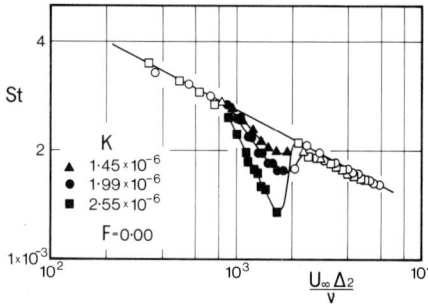

Fig. 19   The effects of acceleration on Stanton
             number with no transpiration.

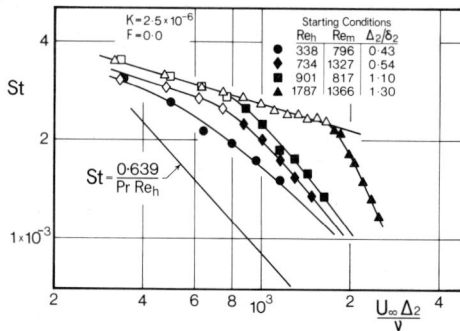

Fig. 20   The effects of initial boundary layer
             thickness on Stanton number during strong
             accelerations with no transpiration.

Reynolds number, and the data points move sequentially
to the left, at constant Stanton number.

     No attempt has been made to devise an empirical
formulation for predicting Stanton number in terms of
enthalpy-thickness Reynolds number, K and F. Complex as the interrelationship is seen to be from these
figures, the experiments cover only a limited range of
conditions.   All the data in these four figures are
from asymptotic accelerated flows, where the flow en-

tered the accelerating region at, or nearly at, the
appropriate value of momentum thickness Reynolds number.   As will shortly be seen, "overshot" or "under-
shot" layers, where the entering values are either
larger or smaller than the asymptotic values, behave
much differently in the accelerating region.

Figures 19 and 20 illustrate the effects of inlet
conditions on the response of the boundary layer to a
strong acceleration.   In Fig. 19, the accelerations
began at momentum and enthalpy thicknesses between 800
and 1000, with the high K runs beginning at the
lower values.   Looking ahead to Fig. 22 shows these
to be nearly the asymptotic values.   The solid symbols in Fig. 19 show the behaviour of the Stanton number within the accelerated region, and display a regular progression of slopes.   With these curves as a
base line, a series of tests was run at a fixed value
of $K = 2.55 \times 10^{-6}$ varying the initial momentum and
enthalpy thickness Reynolds numbers.   The results, in
Fig. 20, show that the slope of the Stanton number correlation is not a unique function of K but depends
upon the initial conditions.   In Fig. 20, the square
symbols represent a near-equilibrium combination, with
momentum thickness and enthalpy thickness Reynolds numbers within 100 units of one another, and of the same
approximate values as shown in Fig. 19.   It is worth
noting that if the enthalpy thickness is kept small
(in this case by delaying the heating) the response
of the Stanton number to the acceleration is diminished.
On the other hand, if the enthalpy thickness is held
nearly constant, and the momentum thickness increased,
there is relatively less change in behaviour from the
reference case.   When both the enthalpy and momentum

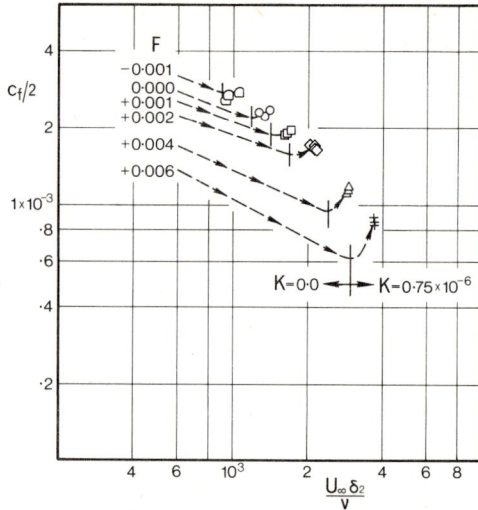

Fig. 21   The behaviour of friction factor and momen-
             tum thickness Reynolds number after in-
             itiation of an asymptotic acceleration con-
             dition.

thicknesses are increased to large values prior to the
acceleration (an "overshot" case), then the Stanton
number comes down very abruptly in the accelerating
region.   For the strongly "overshot" entrance conditions it is not possible to obtain a long run at
equilibrium conditions, hence most of the data shown
are in the region where the boundary layer is still
adjusting to the acceleration.

    The solid line shown for comparison represents
the similarity solution for a laminar wedge flow with
a very thick thermal boundary layer.   It seems clear,
from these data, that the same relative variation of
Stanton number could be attained by different non-
equilibrium combinations of enthalpy-thickness Reynolds
number and acceleration.   The effects of non-equili-

brium combination of momentum thickness and acceleration cannot be uniquely identified by the value of
K alone.   Heat transfer in non-equilibrium acceleration is inherently responsive to all three variables:
the acceleration parameter and the momentum- and
enthalpy-thickness Reynolds number.   Indeed, what
emerges from a study of these figures is that a knowledge of the local values of these parameters is insufficient to determine the local value of St.
"Historical" influences are of great influence.   It is
worth mentioning, however, that while the heat transfer behaviour cannot be adequately correlated with
algebraic formulae, a relatively simple model of
momentum and heat transport (outlined in Sec. 4) manages to account for all the nuances observed in the
experiments.

3.9   *Momentum Transfer to an Accelerated Flow*

The decision to test equilibrium accelerated flows
places some constraints on the behaviour of the momentum boundary layer as illustrated in Fig. 21.   Here,
for accelerations at a value of $K = 0.75 \times 10^{-6}$,   are
trajectories of the boundary-layer behaviour for different values of F.   The broken lines suggest the
behaviour of the skin friction and momentum thickness
Reynolds number at various stations along the plate
prior to the beginning of the acceleration.   The vertical bar shown for each set is the last point in the
unaccelerated flow.   Considering the data for
F = 0.006, after the acceleration begins, the momentum
thickness Reynolds number grows only slightly, from
3000 to 3700, and does not change further with length
along the plate.   At the same time, the value of the

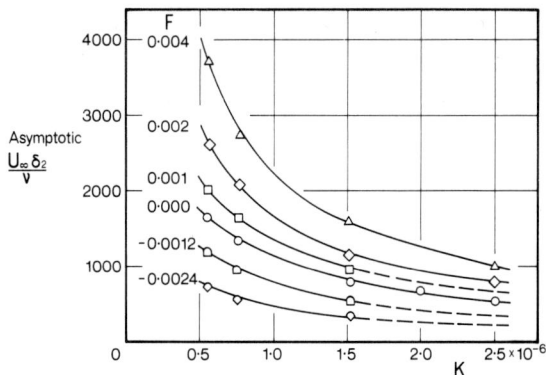

Fig. 22   The asymptotic values of momentum thickness Reynolds number for various values of K and F.

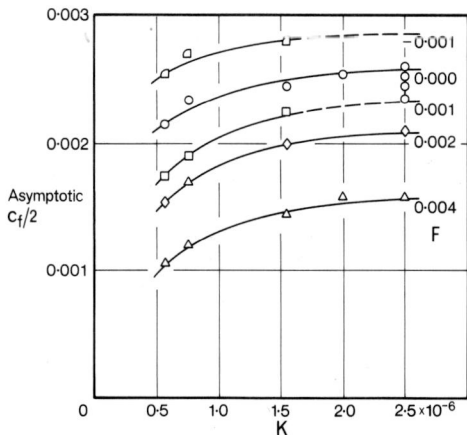

Fig. 23   The asymptotic values of friction factor, $c_f/2$, for various values of K and F.

skin friction coefficient rises quickly to a final value above the flat plate value and then remains unchanged. The equilibrium point thus established is characteristic of this combination of acceleration and blowing. In each of these data sets, the acceleration was begun at or near the predicted value of the equilibrium momentum thickness Reynolds number, to reduce the transient effects as much as possible. The way in which the asymptotic values of momentum-thickness Reynolds number vary with K and F is shown in Fig. 22. Each symbol shown represents an experimentally achieved equilibrium state. Asymptotic values of $c_f/2$ established by these equilibrium flows are shown in Fig. 23. Again, each symbol represents an experimentally achieved equilibrium state. Some confusion existed in the data sets for $K = 2.5 \times 10^{-6}$ and $F = 0$ and four different terminal states were achieved. All are shown, but symmetry with the other data sets suggests that the highest value be used.

The momentum boundary layer for an asymptotic accelerated flow has thus a relatively simple description. Being uniquely specified by F and K, there is no need for a "size dependence" and, in essence, the complexity of description is reduced by one variable. The asymptotic value of friction factor with blowing, can be predicted with reasonable accuracy by applying Eq. (26) to the unblown asymptotic value at the same K. Since F and K uniquely determine the asymptotic thickness, (see Figure 21 or 22) there is no need for a statement "... at the same momentum thickness Reynolds number..." and, in fact, such a proscription cannot be enforced in the context of comparing asymptotic boundary layers with the same value of K and different values of F.

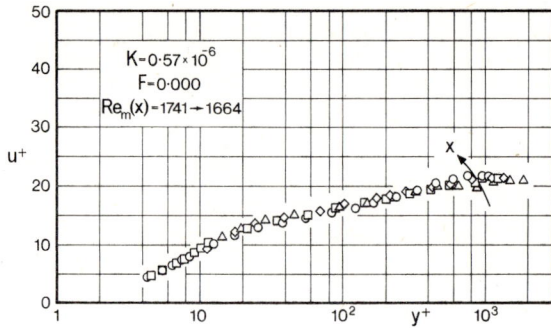

Fig. 24(a)   Sequential velocity profiles within an equilibrium acceleration at K = 0.57×10^{-6} with F = 0.00.

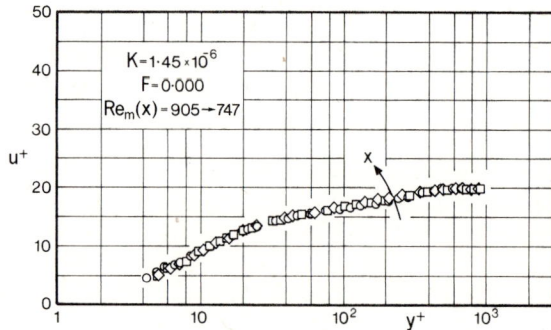

Fig. 24(b)   Sequential velocity profiles within an equilibrium acceleration at K = 1.45×10^{-6} with F = 0.00.

3.10 Mean Velocity and Temperature Profiles in an Accelerating Flow

In a constant-K, asymptotic boundary layer the momentum thickness Reynolds number seeks some characteristic level, as does the friction factor, and the velocity profile assumes a stationary shape in $u^+ \sim y^+$ coordinates.   This is illustrated in Figures 24(a),

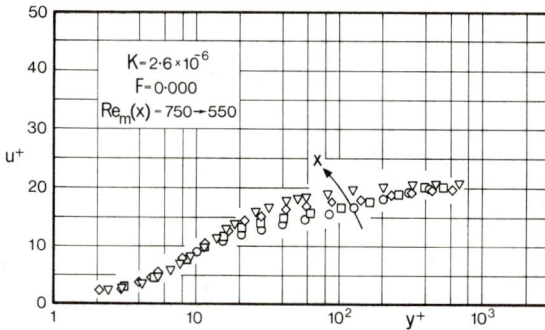

Fig. 24(c)   Sequential velocity profiles within a strong acceleration ($K = 2.6 \times 10^{-6}$) with significant changes in momentum thickness Reynolds number and a large reduction in Stanton number.

24(b) and 24(c) which show the profiles as they developed in the streamwise direction. For the two lower values of K, the momentum thickness Reynolds number was a constant throughout the length of the test section to within 10% and close to the asymptotic values shown in Fig. 22. The profiles show a close similarity in both inner and outer regions. At $K = 2.6 \times 10^{-6}$ the boundary layer was "overshot", entering with a momentum thickness Reynolds number of 750 compared with the asymptotic value of 480 (from Fig. 22). As can be seen, the boundary layer did not reach an asymptotic state, though the last profile (in the streamwise direction) could be taken as representative. The corresponding temperature profiles are shown in Figures 25(a), 25(b) and 25(c). For the two lower accelerations the temperature profiles remain reasonably similar, showing small changes in the outer region (for $y^+ > 100$) which can be seen to increase in magnitude as K increases. Reviewing Figures 15

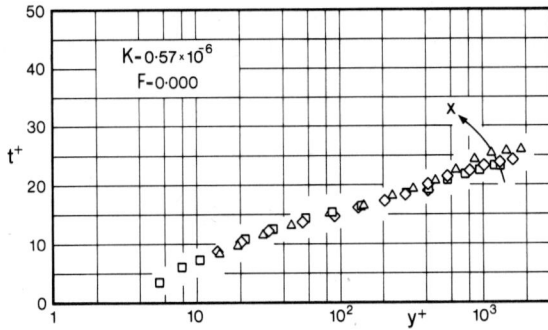

Fig. 25(a)   Sequential temperature profiles within an
              equilibrium acceleration at K = 0.57×10⁻⁶
              with F = 0.00.

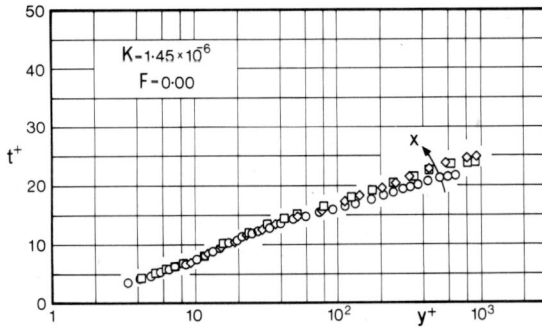

Fig. 25(b)   Sequential temperature profiles within an
              equilibrium acceleration at K = 1.45×10⁻⁶
              with F = 0.00.

and 17 shows that the Stanton-number values for these
conditions were only slightly affected by the acceleration.   When the value of K reaches 2.5×10^{-6},
however, as shown in Fig. 25(c), the temperature profile shows a drastic difference, with the profiles
strongly non-similar in the streamwise direction.   The
effect is felt all the way in to y^+ near 10.   The

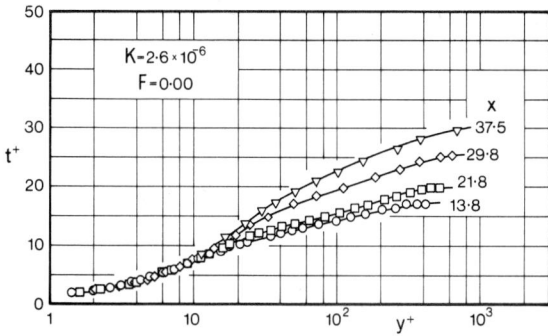

Fig. 25(c)   Sequential temperature profiles within a strong acceleration ($K = 2.6 \times 10^{-6}$) with significant changes in momentum thickness Reynolds number and a large reduction in Stanton number.

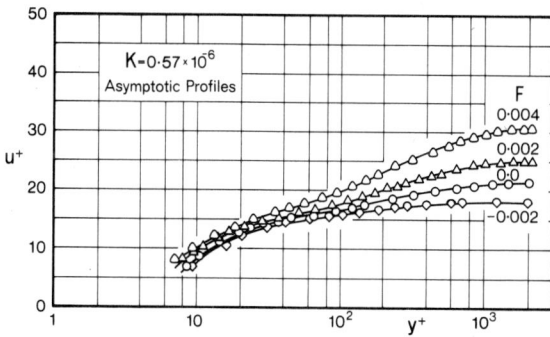

Fig. 26(a)   Asymptotic velocity profiles for mild acceleration with transpiration.

Stanton-number data in Fig. 18 show this combination of conditions to result in a drop in Stanton number which reaches 40% at the downstream end of the test section.

The "terminal states" of the velocity and temperature profiles are shown for high and low accelerations

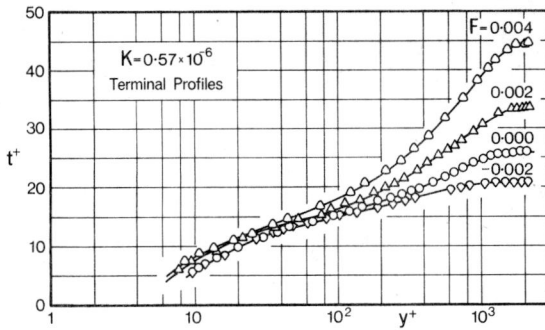

Fig. 26(b)   Terminal temperature profiles observed
              for mild acceleration with transpiration.

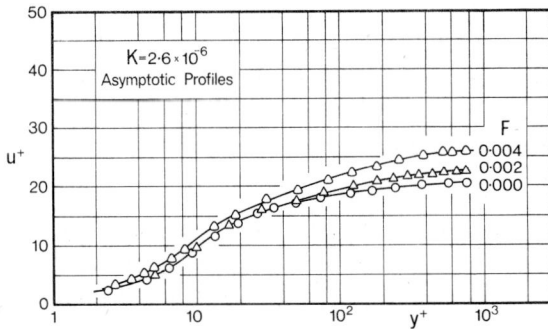

Fig. 27(a)   Asymptotic velocity profiles for strong
              acceleration with transpiration.

($K = 0.57 \times 10^{-6}$ and 2.6×10^{-6}) in Figures 26(a)
and 26(b) and 27(a) and 27(b) for various values of
blowing.   The phrase "terminal states" is used because, while the profiles shown for velocity are representative asymptotic profiles, those shown for temperature are simply the last measured profiles: the
energy boundary layer continues to grow, a longer test
section would have yielded a different "last" profile.

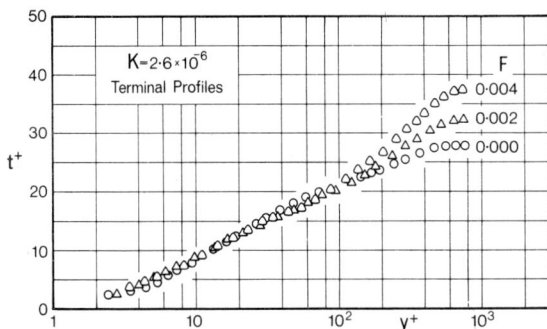

Fig. 27(b)   Terminal temperature profiles observed for strong acceleration with transpiration.

The temperature profiles shown for $K = 2.6 \times 10^{-6}$ display inner region similarity, out to about y^+ of 100; no such coherence is visible in the data for $K = 0.57 \times 10^{-6}$.

The velocity and temperature profiles shown in Figures 24-27 illustrate the main structural features of the accelerated turbulent boundary layer. These data have been used as guides in refining the physical model outlined in Sec. 4 used in the Stanford finite-difference computer program for boundary-layer calculations.

3.11 Flows Subject to Deceleration

Decelerating flows differ from accelerating ones in that no asymptotic boundary-layer state is approached, even though an equilibrium flow is established. The condition of equilibrium between the pressure-gradient force and the shear force is expressed by the parameter β given earlier (Eq. (6)); it has been shown experimentally that boundary layers for which β is uniform

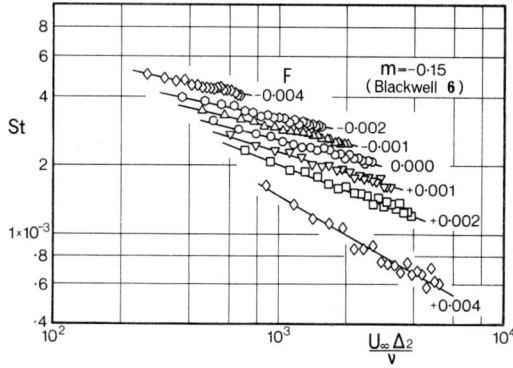

Fig. 28   The variation of Stanton number with en-
             thalpy thickness Reynolds number for mild
             deceleration, with transpiration.

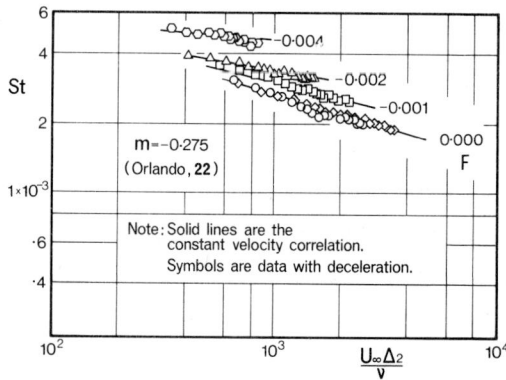

Fig. 29   The variation of Stanton number with en-
             thalpy thickness Reynolds number for strong
             deceleration, with transpiration.

display a constant value of G, the Clauser shape
factor.   Figure 3 has shown that the present data set
indicate the more general result that G remains uniform whenever $(\beta + B_m)$ remains constant, where B_m
is the momentum blowing parameter.

    In Sec. 2 it was shown that the experimental bound-

ary conditions which produce flows with constant β
are those for which the free-stream velocity varies
with distance to some power, m.   This introduces an
experimental difficulty centering around the identification of the virtual origin of the boundary layer.
It would not be appropriate simply to measure "x"
from the leading edge of the test section unless (a)
the boundary layer were of zero thickness at that
point and (b) the boundary layer were fully turbulent
from the leading edge onward.   Andersen [5] observed
that when "x" distance is measured from the virtual
origin (based on extrapolating the turbulent boundary
layer to zero thickness) and velocity varies with some
power of x,   then equilibrium boundary layers are
achieved: both β   and   G   remain substantially constant with length along the test section, after a
brief accommodation.   The decision was also made, for
the whole programme of research, to restrict the study
to flows which did not approach separation.   For this
reason only small negative values of m   were used.
It was anticipated that high blowing would tend to en-
courage separation, hence data were taken only for
cases of suction and small blowing.

    Since no asymptotic state is attained in a decel-
erating equilibrium flow, the Stanton number and friction factor values vary with boundary-layer-thickness
Reynolds numbers and the data resemble the flat plate
data in their general dependences.

3.12 *Heat Transfer in Decelerating Flows*

    Figures 28 and 29 show the variation of Stanton
number with enthalpy-thickness Reynolds number for
moderate and strong decelerations with blowing and

suction.   The solid lines through the data represent
flat plate behaviour.   Evidently, the same correlation
applies to both cases of decelerated flow as applies
for the flat-plate case.   The effects of blowing are
to reduce Stanton number but, again, exactly as was
observed for the flat plate.

Thus, in terms of the surface-heat-transfer behaviour of the boundary layer, one can say that adverse
pressure gradients pose no new problems, within the
range of conditions encountered in this study.   Whatever effects the adverse pressure gradient may have
on the structure of the boundary layer, in terms of
changing the sublayer thickness or the turbulent transport mechanisms, the net effect is the same as for a
flat-plate situation.   Whatever internal correlations
are proposed to describe the effects of pressure gradient on the boundary layer must therefore produce this
same behaviour for decelerating flows.

The heat-transfer characteristics for the decelerated flows reported here can be described by

$$St_o = 0.015 \, Re_h^{-0.25} \tag{34}$$

$$\left. \frac{St}{St_o} \right|_{Re_h} = \left[\frac{ln(1 + B_h)}{B_h} \right]^{1.25} (1 + B_h)^{0.25} \tag{35}$$

Here, St_o is defined as the value of Stanton number
without blowing but in the adverse pressure gradient,
and St is the value of Stanton number in that same
adverse pressure gradient, with blowing, at the same
enthalpy-thickness Reynolds number.

3.13 Momentum Transfer in Decelerating Flows

Although the heat-transfer behaviour in decelerating flows can be adequately described in terms of
flat-plate correlations, momentum transfer cannot.
The effect of an adverse pressure gradient is to de-
celerate the fluid in the boundary layer causing the
momentum thickness to increase more rapidly than it
would due to wall shear alone.   The variation of
$c_f/2$ with pressure gradient in the absence of transpiration is shown in Fig. 30 and summarized by the
following recommendations, for flows in which
$U_\infty = U_1 x^m$ with m a constant.

$$\left(\frac{c_{f_0}}{2}\right) = a\ Re_m^{-0.25} \tag{36}$$

where

U_∞ = constant	a = 0.0120	$(850 < Re_m < 3000)$
$U_\infty = U_1 x^{-.15}$	a = 0.0102	$(1500 < Re_m < 3500)$
$U_\infty = U_1 x^{-.20}$	a = 0.0083	$(1700 < Re_m < 4200)$
$U_\infty = U_1 x^{-.275}$	a = 0.0059	$(2000 < Re_m < 5000)$

Near the entrance of the test section the values of
G and β were usually not stabilized, with β continuing to rise for the first two or three data points
to its final value.   The forms of the curves therefore
reflect this accommodation: only the last six data
points should be taken to represent equilibrium conditions.   The data for m = -0.275 show much scatter:
these are difficult conditions under which to measure
the friction factor.   The line shown passing through
the data must be regarded as only a suggestion, at
best.

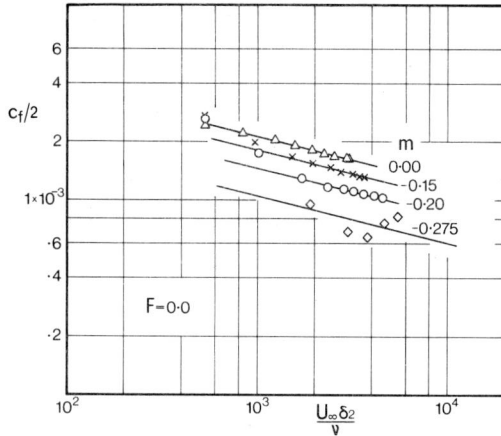

Fig. 30   The variations of friction factor, $c_f/2$, with momentum thickness Reynolds number for three decelerating flows.

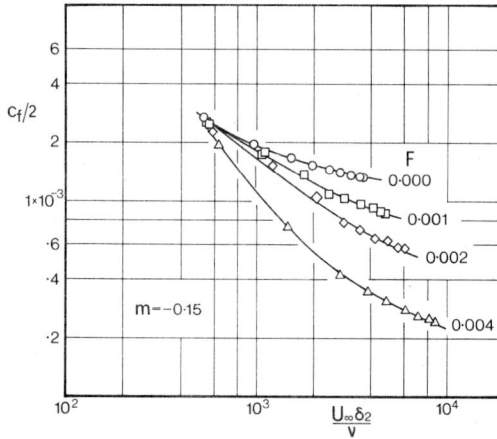

Fig. 31   The variation of friction factor with momentum thickness Reynolds number for mild deceleration with transpiration.

Blowing increases the momentum deficit of the boundary layer and would aggravate a tendency to sep-

arate due to an adverse pressure gradient, so combinations of blowing with deceleration are prone to
early separation.   For this reason, the blowing data
shown on Fig. 31 are restricted to a moderate deceleration   (m = -0.15)   and moderate blowing.   The data
in Fig. 31 include all recorded values within the region of constant   m.   In the entrance region the
values of β and   G   were not usually stabilized,
with β continuing to rise for the first two or three
data points.   In this region of increasing β,   the
value of $c_f/2$ shows a rapid drop with Reynolds number.   In the region where β was substantially uniform, the variation of $c_f/2$ with Reynolds number is
similar to that observed on a flat plate.   In particular, the relative effects of blowing on skin friction
are similar to those observed on a flat plate (though,
of course, the unblown values are much different) and
are once again predicted well by:

$$\left.\frac{c_f}{c_{f_o}}\right|_{Re_m} = \left[\frac{ln(1 + B_m)}{B_m}\right]^{1.25} (1 + B_m)^{0.25} \tag{37}$$

where B_m is the momentum blowing parameter.

3.14 Velocity and Temperature Profiles for Decelerating Flows

Velocity and temperature profiles for moderate
deceleration   (m = -0.15)   are shown in Figures 32 and
33 in wall coordinates.   The velocity profile is relatively unaffected by the deceleration for $y^+ < 200$
either for suction or for no blowing.   Blowing at
0.004 has a very pronounced effect, however, raising
the values of u^+ at every y^+ greater than about 10.

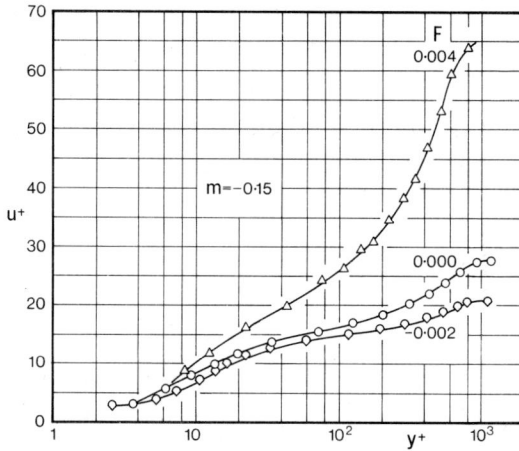

Fig. 32   Velocity profiles for mild deceleration
           with transpiration.

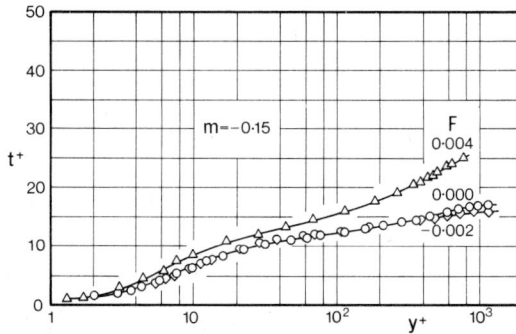

Fig. 33   Temperature profiles for mild deceleration
           with transpiration.

The profiles of t^+ are less affected by the deceleration than are the profiles of u^+, being slightly
lower across the board.

3.15 Integral Relationships for Friction Factor, Stanton Number and Shape Factor

The preceding sections have discussed data for flat-plate flows (U_∞ = constant), asymptotic accelerating flows (K = constant) and equilibrium decelerating flows (m = constant). For each of these hydrodynamic situations data have been presented for different transpiration levels both positive (blowing) and negative (suction). In addition, one can envisage flows with wall temperature variations such that the thermal and momentum boundary layers might be of considerably different thicknesses. The number of combinations of these boundary conditions is very large, and it would indeed be surprising if any correlation could be contrived which would describe Stanton-number or friction behaviour directly and which would cover more than a small range of these conditions with any accuracy. Generality of the sort needed for that task seems inherently to require a differential predictor scheme with the experimental inputs providing information about the transport mechanisms within the boundary layer. Such is the approach taken by most heat-transfer research today. The differential correlations needed to describe these experimental data are discussed in Sec. 4 and their success in the prediction of complex combinations is shown in Sec. 4.3.

For the present, however, it can be said that some correlations can be given which are useful for situations not too far removed from the equilibrium states represented by these data. In particular, situations involving slowly varying blowing, or slowly varying K or β even for a relatively large difference between

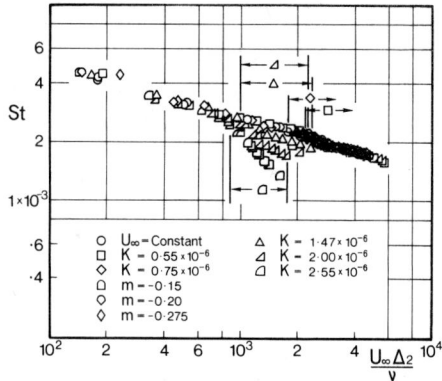

Fig. 34   A summary of Stanton number behaviour for all flows tested.

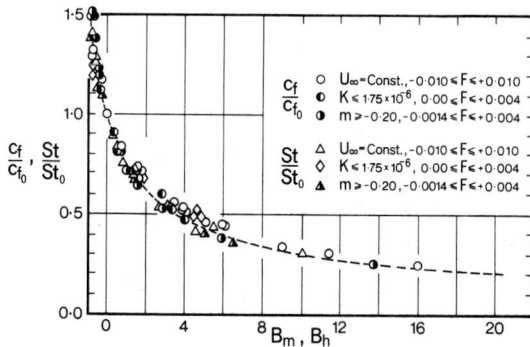

Fig. 35 ·   A summary of the effects of transpiration on Stanton number and friction factor for all flows tested.

the thermal and momentum boundary-layer thicknesses.

Such correlations are shown in Figures 34-36: Stanton number versus enthalpy thickness, the effects of blowing, and the variation of shape factor.   In choosing the correlations, conflicts had to be re-solved between the desire for accuracy and the desire for range.   The correlations shown are believed valid

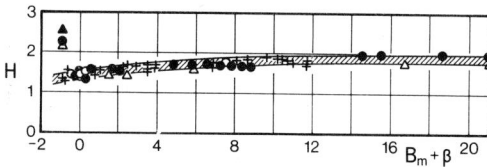

Fig. 36   The variation of shape factor,   H,   with
            $B_m + \beta$ for all flows tested.

within ±10% for the ranges of values covered. Figure
34 shows Stanton number versus enthalpy thickness
Reynolds number for a number of different cases, all
without transpiration: flat plate, moderate and
strong accelerations, and moderate to strong decelerations.   All of the data shown are from the present
series, and represent equilibrium states.   It is
noteworthy that a single correlation covers the data
for all flat plate flows, all decelerating flows, and
all accelerating flows less severe than $K = 1.47 \times 10^{-6}$.
The recommended curve is:

$$St_o = 0.015 \; Re_h^{-0.25} \qquad (AIR) \qquad (38)$$

$$150 < Re_h < 6000$$

In recommending this particular form it must be ad-
mitted that certain traditions have been honoured.   In
particular, the use of the exponent -0.25 tends to
weight more heavily the data from the high Reynolds
number region; but this value has a long analytic
history.   The skin-friction data are not so conveniently described but the flat plate and the asymptotic states of accelerated flows up to $K = 2.5 \times 10^{-6}$
are reasonably well described in terms only of the
momentum thickness Reynolds number by

$$\frac{c_{f_o}}{2} = 0.0128 \ Re_m^{-.25} \qquad \begin{array}{l} 0 \le K \le 2.5 \times 10^{-6} \\[6pt] (AIR) \\[6pt] 500 \le Re_m < 5000 \end{array} \qquad (39)$$

It should be borne in mind that for each asymptotic accelerated flow there exists only one possible value for Re_m, dependent upon the value of K: the asymptotic value, given earlier in Fig. 22. When used to predict these asymptotic values of $c_f/2$ from the asymptotic values of Re_m, the equation given above tends to underpredict by about 10%.

Blowing can be discussed either in terms of the effects at a particular location (x-Reynolds number) or at a particular local state of the boundary layer (enthalpy thickness Reynolds number). The local descriptor has more generality since it can be applied in cases of non-uniform velocity. It has been found, by comparison with the data presented here and in the original source documents, that the effect of blowing (or suction) can be calculated with reasonably good accuracy using a form derivable from a Couette-flow model, evaluated at constant boundary-layer-thickness Reynolds number:

$$\left. \frac{c_f}{c_{f_o}} \ , \ \frac{St}{St_o} \right|_{\substack{Re_m \\ \text{or} \\ Re_h}} \simeq \left[\frac{ln(1 + B)}{B} \right]^{1.25} (1 + B)^{0.25} \qquad (40)$$

where B refers either to the momentum- or enthalpy-based blowing parameter as appropriate.
This relationship is recommended within the following

range of conditions.

U_∞ = constant flows (-0.01 F ≤ 0.010)

Decelerating flows (-0.20 ≤ m;

-0.004 ≤ F ≤ +0.004)

Accelerating asymptotic flows (K ≤ 1.75 × 10^{-6};

0 ≤ F ≤ +0.004)

The range of applicability of Eq. (40) can be recalled by the following notation, which suggests that three parameters are important:

$$\frac{c_f}{c_{f_o}} \simeq \frac{St}{St_o} = f(U_\infty(x); \quad B; \quad size) \qquad (41)$$

For accelerating flows with constant K an asymptotic condition may be reached such that "size" is a unique function of K and B.   For such conditions the list of variables is reduced to two, since "size" is fixed once K and B are chosen.   Hence the comparison can be made between asymptotic states.   For flat plate and decelerating flows, "size" is a variable and the comparisons must be made at the same boundary layer thickness Reynolds numbers.   Confirmation of the validity of Eq. (40) is shown in Fig. 35 which includes friction and heat transfer data for accelerating, decelerating, and flat plate cases, with blowing and with suction.

The velocity-profile shape factor, H, has been found to correlate reasonably well for all values of blowing and suction and for all variations of free-stream velocity if described in terms of $(B_m + \beta)$, a parametric group occurring in one form of the momentum integral equation.   Figure 36 shows H versus $(B_m + \beta)$ for conditions covering accelerations and decelerations with uniform values of blowing along the

surface.   With the exception of 4 data points at
$(B_m + \beta) = -1.0$,   the remainder of the data are well
organized.   The errant points may well have been
laminarized by the combined effects of suction and
acceleration.

4.   A MATHEMATICAL MODEL FOR PREDICTING THE BOUNDARY-
     LAYER BEHAVIOUR

4.1   Solution of the Momentum Equation

During the past decade enormous strides have been
made in our ability to solve the partial differential
equations of the boundary layer, using finite-difference
techniques and the power of the digital computer.   To
all intents and purposes mathematically exact solutions
to the boundary-layer equations can be obtained for
virtually any kind of boundary conditions, provided
that the turbulent transport processes are adequately
modelled.   The speed with which such solutions can be
obtained has made direct solution of the boundary-layer
equations for particular applications a practical engineering design tool.

We are not going to be concerned here with the
details of any of the several finite difference procedures in common use today, but rather with a scheme
that has been used successfully to model the dominant
turbulent shear stresses in transpired turbulent boundary layers.

The time averaged momentum equation of the boundary layer, particularised for the moment to constant
fluid properties and neglecting turbulent normal
stresses, may be written as follows:

$$u \frac{\partial u}{\partial x} + v \frac{\partial u}{\partial y} - \frac{\partial}{\partial y} \left[\nu \frac{\partial u}{\partial y} - \overline{u'v'} \right] + \frac{1}{\rho} \frac{dp}{dx} = 0 \qquad (42)$$

If the turbulent shear stress $\overline{u'v'}$ is known at all points in the boundary layer, the velocity distribution may be found simply by solving Eq. (42) and the continuity equation for any desired boundary conditions, including transpiration.

Although progress continues to be made in turbulent transport theory in general, and turbulent boundary-layer theory in particular, it is still fair to say that there is as yet no truly fundamental theory that may be used as a universal starting point for solution of turbulence problems. Turbulent boundary-layer theory has gone through, and continues to go through, a series of stages involving successively higher orders of sophistication. Each level in this development has involved the correlation of experimental data at a more fundamental level, and has opened up the possibility for solving successively broader ranges of problems with a single consistent set of empirical constants. The information and calculating procedures to be presented here do not represent any very bold steps toward a more general theory, but they will allow computation of equilibrium and near-equilibrium boundary layers as precisely as any scheme so far devised. Higher order models are presently being investigated by numerous researchers, and hopefully will lead to theories that embrace still broader classes of applications, although probably at the price of complexity and computation cost.

We will first introduce the concept of eddy diffusivity for momentum, ε_m, as a convenient way of expressing the turbulent shear stress.

$$\overline{u'v'} = -\varepsilon_m \frac{\partial \overline{u}}{\partial y} \qquad (43)$$

Already we are in the realm of theoretical contro-
versy since implicit in Eq. (43) is the notion that
the turbulent shear stress vanishes in the absence of
a gradient in mean velocity.   In spite of its short-
comings, the eddy diffusivity concept has the virtue
of allowing one to use the same computational program
for both laminar and turbulent boundary layers.   Since
most real turbulent boundary layers grow out of laminar
boundary layers, the advantage is obvious.

It is convenient to visualize the turbulent boundary layer as consisting of an inner wall-dominated
region, and an outer region which actually extends over
most of the boundary layer.   However, for most applications the inner region is, by far, the more important
one, and it is on this region that we will focus most
attention.

The inner region may be subdivided into a zone
immediately adjacent to the wall in which viscous forces
predominate (ε_m approaches zero), and a region farther
out in which momentum transfer is almost entirely by
turbulent transport processes, but in which the scale
and intensity of the turbulence is still strongly dependent upon the proximity of the wall.

The Prandtl mixing-length theory, despite much
criticism for many years, still provides a simple and
remarkably adequate basis for describing the turbulent
momentum transport process in the inner region, at
least for equilibrium and near-equilibrium boundary
layers.   The mixing length, ℓ, is defined such that

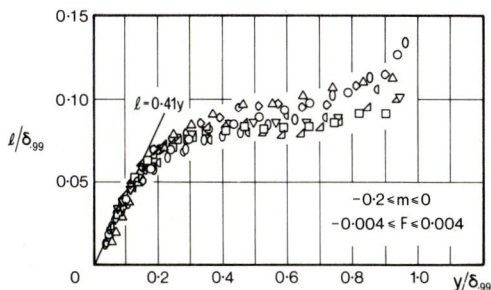

Fig. 37   The behaviour of the mixing length for flat plate and decelerating flows with transpiration.

it is related to the eddy diffusivity for momentum and the mean velocity gradient by the following equation:

$$\varepsilon_m = \ell^2 \left| \frac{\partial u}{\partial y} \right| \tag{44}$$

Outside of the viscous-dominated region immediately adjacent to the wall, the mixing-length in the inner part of the boundary layer is found to be proportional to distance from the wall, with a proportionality factor, k, that is independent of either transpiration rate or pressure gradient.   Figure 37 shows some measurements of the mixing-length for a number of cases of transpiration, both blowing and suction, with no pressure gradient and with an adverse pressure gradient.   Results for favourable pressure gradients are similar.   Note that all the data in the region near the wall converge on a single linear relation with k ≃ 0.41.   We will model the region outside the viscous near-wall zone (which we will now term the viscous sublayer), but inside of the outer, or "wake", region, by:

$$\ell \;=\; ky \qquad\qquad (45)$$

where $k \;=\; 0.41$

The viscous sublayer immediately adjacent to the wall can be modelled in a simple way by introducing a damping function that causes mixing length ℓ to fall to zero as the wall is approached faster than Eq. (45) suggests. Designating the damping function as D, the mixing length over the entire inner region may then be expressed as:

$$\ell \;=\; kyD \qquad\qquad (46)$$

The damping function D can be satisfactorily expressed in a number of different ways. A scheme which was first suggested in 1956 by Van Driest [15], yet which remains popular today, is an exponential function which leads to mean velocity profiles that correspond quite well with those observed experimentally

$$D \;=\; 1.0 - \exp(-y^{+}/A^{+}) \qquad\qquad (47)$$

where A^{+} may be interpreted as the effective thickness of the viscous sublayer expressed in terms of inner-layer coordinates.

The effective thickness of the viscous sublayer is probably the single most important parameter in the computation of turbulent boundary layers. The sublayer, though comprising a very small fraction of the total boundary layer thickness, is the region where the major change in velocity takes place, and, except for very low Prandtl number fluids, is the region wherein most of the resistance to heat transfer resides. If this region is modelled accurately one may usually

escape with a relatively crude approximation throughout the rest of the boundary layer.

The thickness of the sublayer is evidently determined by viscous stability considerations. The experimental evidence is that a favourable pressure
gradient (dp/dx negative) results in increased thickness, while an adverse pressure gradient has the opposite effect. Transpiration into the boundary layer
(blowing) decreases the thickness, if it is expressed
in non-dimensional wall coordinates, while suction has
the opposite effect. Surface roughness, while not a
subject of this article, causes a thinning of the sublayer.

The effects of pressure gradient and transpiration
on A^+ are conveniently expressed in terms of a non-
dimensional pressure-gradient parameter, p^+, defined
as $(\nu/\rho U_\tau^3)dp/dx$, and a non-dimensional blowing parameter, v_o/U_τ, which is abbreviated as v_o^+.

The functional dependence of A^+ upon p^+ and
v_o^+ has been deduced experimentally by examination of
a very large number of velocity profiles obtained as
part of the Stanford project over a period of six years.
Before examining these results, however, it should be
mentioned that considerable progress has been made in
both qualitatively and quantitatively describing this
function using some relatively simple theoretical ideas
and a minimum of experimental data. A number of investigators (for example, Bradshaw [16]) have discussed
the significance of a minimum value of a local Reynolds
number of turbulence as being a requisite for mainten-
ance of a turbulent boundary layer. Numerous investigators, going back to the early theoretical work on
the transpired boundary layer by Rubesin [17], have

implicitly introduced this concept as a basis for de-
fining the thickness of the viscous sublayer.   The
local Reynolds number of turbulence can be defined as:

$$Re_t \;=\; \ell_t \sqrt{|\overline{u'v'}|} \Big/ \nu \;=\; \ell_t \sqrt{\tau_t/\rho} \Big/ \nu \qquad (48)$$

ℓ_t, the turbulence length scale, can be considered
to be effectively the same as the mixing-length
ℓ = ky.   Thus

$$Re_t \;=\; ky \sqrt{\tau_t/\rho} \Big/ \nu \qquad (49)$$

It should also be noted that by combining Equations
(43) and (44) with (48) and (49),

$$Re_t \;=\; \varepsilon_m/\nu \qquad (50)$$

Furthermore, outside of the viscous sublayer $\tau_t \sim \tau$,
so Eq. (49) can also be expressed as,

$$Re_t \;=\; ky^+ \sqrt{\tau/\tau_o} \qquad (51)$$

An examination by Andersen [5] of a large amount
of experimental data for transpired turbulent boundary
layers for both favourable and adverse pressure gradients indicates that Re_t is approximately the same
number (about 33.0) in every case at a point outside
of the sublayer defined as approximately $y^+ = 2.5\ A^+$.
Thus the thickness of the viscous sublayer, and by implication A^+, is evidently characterized by a critical value of the Reynolds number of turbulence.   It
follows, incidentally, that if Re_t falls everywhere
below this value the turbulence in the boundary layer
will damp out and a laminar boundary layer will result,
and this is precisely what is observed in strongly ac-

celerated flows where the shear stress decreases so
rapidly with distance from the wall that Re_t never
reaches 33.0.   The important point here, however, is
that with these facts alone it is possible to generate
the functional dependence of A^+ upon p^+ and v_o^+.

The following equation is an empirical representation of the experimental data on A^+, but it could
just as well be described as an empirical representation of Andersen's analysis.   In either case the
algebraic form of the equation has no particular significance.

$$A^+ = \frac{24.0}{a\left[v_o^+ + b\left(\dfrac{p^+}{1 + cv_o^+}\right)\right] + 1.0} \tag{52}$$

where $a = 7.1$ if $v_o^+ \geq 0.0$, otherwise $a = 9.0$

$b = 4.25$ if $p^+ \leq 0.0$, otherwise $b = 2.0$

$c = 10.0$ if $p^+ \leq 0.0$, otherwise $c = 0.0$

Equation (52) is plotted on Fig. 38 where the
effects of pressure gradient and transpiration can be
clearly seen.   Note that a strong favourable pressure
gradient forces A^+ to very high values, and that
blowing lessens this effect, while suction increases
it.   If A^+ becomes very large the viscous sublayer
simply overwhelms the entire boundary layer;   this is
the "laminarization" discussed earlier.   In fact most
of the trends noted earlier in connection with the experimental data on Stanton number are recoverable by
varying the value of A^+.   The thickening of the sublayer caused by a favourable pressure gradient (accelerating flows) results in a decreased Stanton number
simply because the major resistance to heat transfer

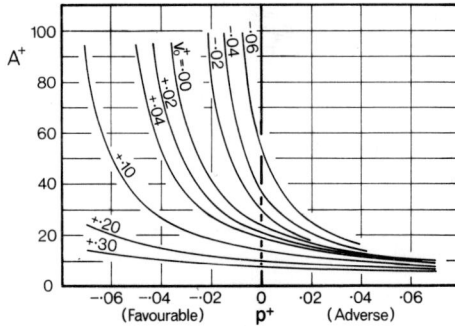

Fig. 38   The variation of the damping constant, A^+, as a function of v_o^+ and p^+.

is in the viscous sublayer.

Note that an adverse pressure gradient causes a decrease in sublayer thickness. Interestingly, when these results are used to compute velocity profiles for adverse pressure gradients without transpiration, and the velocity profiles are plotted on $u^+ \sim y^+$ coordinates, they tend to fall on the same line as is obtained for zero pressure gradients in the near-wall region (but outside of the sublayer). This is the "law-of-the-wall" which has long been known to provide an excellent fit to experimental data for both zero-pressure-gradient and adverse-pressure-gradient flows.* *The universality of the "law-of-the-wall" for adverse pressure gradients thus appears to result from compensating effects of the decreased sublayer thickness and the positive pressure gradient.*

A^+ as represented by Eq. (52) and Fig. 38 has been evaluated under essentially equilibrium conditions, *i.e.*, conditions under which v_o^+ and/or p^+ are in-

----

* The conventional "law-of-the-wall" does not apply for strong favourable pressure gradients.

variant or, at worst, are varying only slowly along
the surface.   This is a case of inner-region equilibrium.   It is probable that when a sudden change of
external conditions is imposed, the inner region comes
to equilibrium more rapidly than the outer region,
although this has not been proved.   In any case, under
non-equilibrium conditions where v_o^+ or p^+ are
changing rapidly, it has been observed that the sublayer does not change instantaneously to its new equilibrium thickness, *i.e.*, A^+ does not immediately
assume its new equilibrium value.   It can be hoped
that some of the higher order models of turbulence
will predict this effect, but in the meantime, a
reasonably satisfactory expedient is to use a rate
equation of a type suggested by Launder and Jones
[18]:

$$\frac{dA_{eff}^+}{dx^+} = (A_{eff}^+ - A_{eq}^+)/C \qquad (53)$$

A_{eff}^+ is the locally effective value of A^+, while
A_{eq}^+ is the equilibrium value obtained from Eq. (52).
A value of C of about 4.0 has been found to be
reasonable.

All the discussion so far has been concerned with
the inner region of the boundary layer.   The outer
region, extending over the outer 75% or so of the
boundary layer, is of considerably less importance in
predicting performance, and thus can be handled successfully using more gross approximations.   This statement may not be valid for strongly non-equilibrium
boundary layers, especially under adverse pressure
gradient conditions.   Its validity for accelerating
flows with and without transpiration will be demon-

strated later.   In any case, for equilibrium or near-
equilibrium boundary layers, either the assumption of
a constant value of eddy diffusivity over the entire
outer region, or the assumption of a constant value
of mixing length over the entire outer region yields
approximately the same result.   If a constant eddy
diffusivity is used, an empirical correlation of eddy
diffusivity as a function of either displacement or
momentum thickness Reynolds number can be obtained.
However, if mixing-length is used in the inner regions,
it is computationally simpler to use the mixing-length
concept for the entire boundary layer.

Figure 37 shows measured mixing-length data for
the outer region for a number of cases of transpiration
with zero and adverse pressure gradients.   The ade-
quacy of an assumption that the mixing length is constant in the outer region may be judged from these
data.   A further simplification is also illustrated
in this figure: the magnitude of the mixing length in
the outer region is approximately proportional to the
total boundary-layer thickness.   A satisfactory computation scheme is to express the outer region mixing-
length as a fixed fraction, λ, of the 99% boundary-
layer thickness.

$$\ell \;\; = \;\; \lambda \; \delta_{99} \tag{54}$$

A value of $\lambda = 0.08$ works reasonably well over the
entire range of experimental data discussed in this
article, including favourable and adverse pressure
gradients, blowing and suction.   One then simply
evaluates ℓ from Eq. (46) until the value obtained
equals that given by Eq. (54), and then uses the latter
value for the remainder of the boundary layer.

There is some evidence that the effective value
of λ is larger than 0.08 for boundary layers in
which the momentum thickness Reynolds numbers is less
than 5500.   This may be a result of the fact that at
low Reynolds numbers the sublayer is a larger fraction
of the boundary layer and the approximation of a constant mixing length over the remainder of the boundary
layer is less valid.   For strong blowing, even at low
Reynolds numbers, λ again appears to be close to
0.08, and this is consistent with the above explanation
because the sublayer is then thinner.   The following
equation has been found to describe the observed low-
Reynolds-number behaviour of λ quite well:

$$\lambda = 0.235 \, \text{Re}_m^{-1/8} \, (1. - 67.5 \, F) \qquad (55)$$

$$\text{IF} \quad \lambda < 0.08; \quad \lambda = 0.08$$

4.2   Solution of the Energy Equation

The time-averaged energy equation of the boundary
layer, particularized to constant fluid properties and
negligible viscous dissipation, and neglecting turbulent conduction in the stream-wise direction, may be
written as:

$$u \, \frac{\partial T}{\partial x} + v \, \frac{\partial T}{\partial y} - \frac{\partial}{\partial y} \left[\alpha \, \frac{\partial T}{\partial y} - \overline{T'v'} \right] = 0 \qquad (56)$$

This equation can be solved for any desired boundary conditions providing the velocity field has been
established first by solution of the momentum equation,
and provided that we have information on the turbulent
heat transfer rate, $\overline{T'v'}$.

Analogous to the method of solution of the momentum equation, we will introduce the concept of eddy

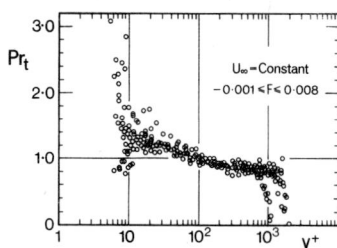

Fig. 39   The variation of turbulent Prandtl number
            within the boundary layer for flat plate
            flows with transpiration.

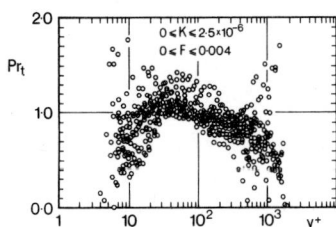

Fig. 40   The variation of turbulent Prandtl number
            within the boundary layer for accelerating
            flows without transpiration.

diffusivity for heat, ε_h.

$$\overline{T'v'} \;=\; -\varepsilon_h \frac{\partial T}{\partial y} \tag{57}$$

Although it might be fruitful to attempt to evalu-
ate either $\overline{T'v'}$ or ε_h on the basis of assumptions
that are independent of the turbulent shear stress,
it seems plausible that there is some kind of relation-
ship between $\overline{T'v'}$ and $\overline{u'v'}$, or ε_h or ε_m. Therefore most analysts have found it convenient to intro-
duce the concept of a turbulent Prandtl number, Pr_t,
defined as follows:

$$Pr_t = \frac{\varepsilon_m}{\varepsilon_h} \tag{58}$$

Introducing Equations (57) and (58) into Eq. (56) we obtain:

$$u\frac{\partial T}{\partial x} + v\frac{\partial T}{\partial y} - \frac{\partial}{\partial y}\left[(\alpha + \varepsilon_m/Pr_t)\frac{\partial T}{\partial y}\right] = 0 \tag{59}$$

If Pr_t were known, Eq. (59) could be solved for any desired boundary conditions so long as the momentum equation had been previously solved. Evaluation of the turbulent Prandtl number can thus solve one of the central problems of turbulent heat transfer.

A very simple physical model of the turbulent momentum and energy transfer processes leads to the conclusion that $\varepsilon_h = \varepsilon_m$, *i.e.*, $Pr_t = 1.00$ (the "Reynolds Analogy"). Slightly more sophisticated models suggest that $Pr_t > 1.00$ when the molecular Prandtl number, Pr, is less than unity. Still other models suggest that Pr_t equals 0.7 or 0.5 in turbulent wakes.

The experimental data are not abundant, but Figures 39, 40, 41, and 42 show the measurements, respectively of Simpson, Whitten and Moffat [19], Kearney [14], and Blackwell [6] with air as a working substance. These were all evaluated from measurements of the slopes of mean velocity and temperature profiles, together with estimates of shear stress and heat flux profiles, and the experimental uncertainty is high, especially near the wall ($y^+ < 20$) and near the outer edge of the boundary layer. The data on Fig. 39 are all for constant free-stream velocity, but cover a wide range of blowing and suction conditions.

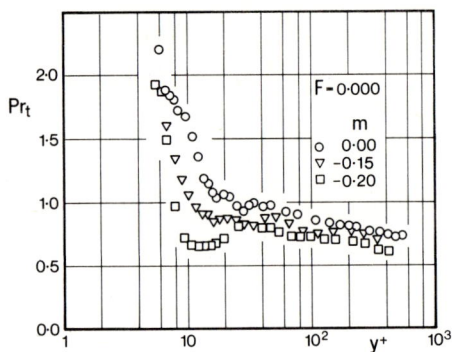

Fig. 41   The variation of turbulent Prandtl number
           within the boundary layer for decelerating
           flows with transpiration.

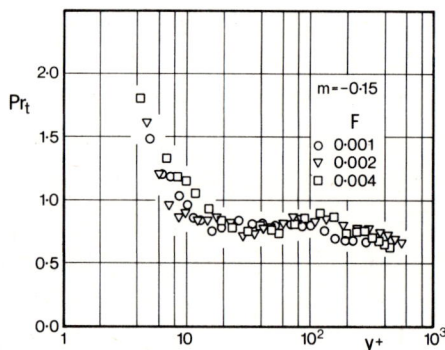

Fig. 42   The variation of turbulent Prandtl number
           within the boundary layer in a mild deceler-
           ation, with transpiration.

The data on Fig. 40 are for accelerated flows with a
considerable range of blowing.   Figure 41 shows three
separate test runs with no transpiration, but first
with no pressure gradient, and then two cases of successively stronger equilibrium adverse pressure gradients.   Finally Fig. 42 shows three test runs for an

adverse pressure gradient with three cases of successively stronger blowing.

Despite the very considerable scatter of data, a
few conclusions seem definitely warranted. First,
the turbulent Prandtl number, at least for air, appar-
ently has an order of magnitude of unity. Thus the
Reynolds Analogy (Pr_t = 1.0) is not a bad approximation.

The second conclusion is that Pr_t seems to go
to a value higher than unity very near the wall, but
is evidently less than unity in the wake or outer
region. The situation very close to the wall is especially vexing because it is extremely difficult to
make accurate measurements in this region, and yet it
seems evident that something interesting and important
is happening in the range of y^+ from 10.0 to 15.0.
The behaviour of Pr_t at values of y^+ less than
about 10.0, is highly uncertain but fortunately not
very important because molecular conduction is the
predominant transfer mechanism in this region. At
the other extreme, in the wake region Pr_t does not
need to be known precisely because the heat flux tends
to be small there.

Another conclusion, for which the evidence is not
yet very strong, is that there is some small effect
of pressure gradient on Pr_t. Figure 41 suggests that
an adverse pressure gradient tends to decrease Pr_t.
Although the scatter of his data was very large,
Kearney [14] reported that there seemed a tendency for
Pr_t to be increased by a favourable pressure gradient
(an accelerating flow).

The results on Fig. 39 suggest that transpiration
does not influence Pr_t unless there is an effect very

close to the wall that is hidden in the experimental uncertainty in this region.   This conclusion is also implied by the results on Fig. 42.

Many analysts have been content to assume that the turbulent Prandtl number is a constant throughout the boundary layer, and indeed the assumption that Pr_t = 0.9, for air, will generally yield satisfactory predictions of overall heat transfer rates.   However, it is found that the assumption of a constant Pr_t will yield temperature profiles that do not correspond well with experiment except in the regions very close to the wall, and near the outer edge of the boundary layer.   Temperature profiles can be much more accurately predicted if some attempt is made to introduce a variation of Pr_t with y^+ that at least approximates the variation seen in the experimental data. Both of the following equations, neither of which have any theoretical basis, have been used with reasonable success by the authors for calculations for air:

$$Pr_t = 1.43 - 0.17 \ y^{+\frac{1}{4}} \tag{60}$$

$$\text{IF} \quad Pr_t < 0.86; \quad Pr_t = 0.86$$

$$Pr_t = 0.90 + 0.35 \ [1 + \cos(\pi y^+/37)]; \quad y^+ < 37 \tag{61}$$

$$= 0.90; \quad y^+ > 37$$

$$= 0.60; \quad y^+ > (\lambda \delta_{99}/k)$$

A pressure gradient effect has not been included in these empirical equations because of insufficient information but, in Sec. 4.3 below, one effect of this omission will be illustrated.

4.3 Some Examples of Boundary Layer Predictions

The quality of boundary-layer predictions that
can be made using the mixing-length model and associated empirical functions will now be demonstrated.
Four examples have been chosen for illustration. The
first is the case of the simple impermeable wall with
no pressure gradient, and this is of course both an
equilibrium momentum boundary layer and an equilibrium
thermal boundary layer. The second is an adverse
pressure gradient equilibrium boundary layer. The
third is an adverse pressure gradient boundary layer
with strong blowing which is not precisely an equilibrium boundary layer, but shows near-equilibrium
characteristics. The final example is a strongly accelerated boundary layer with strong blowing, but in
which both blowing and acceleration are abruptly
stopped at different points along the surface to yield
non-equilibrium conditions.

A modification of the Spalding/Patankar [20]
finite-difference program was used for all predictions,
although any good finite-difference procedure should
yield similar results.

Figure 43 shows $c_f/2$ as a function of momentum
thickness Reynolds number for the simple impermeable
wall with no pressure gradient. Shown for comparison
is the recommendation of Coles [21], which is based on
an extensive examination of the available data, and
also two sets of data from the Stanford project, the
earlier results of Simpson, and the more recent results
of Andersen. The predicted friction coefficients co-
incide closely with Coles' correlation; indeed the
auxiliary functions were chosen to force this coinci-
dence.

0·004

0·003

$c_f/2$

0·002

Simpson **8**

Andersen **5**

Coles **21**

Present prediction

0·001

10^3 1·5   2   3 5×10^3

$\dfrac{U_\infty \delta_2}{\nu}$

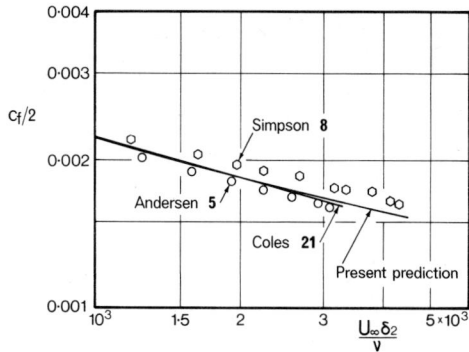

Fig. 43   A comparison of measured and predicted values of friction factor for a flat plate flow without transpiration.

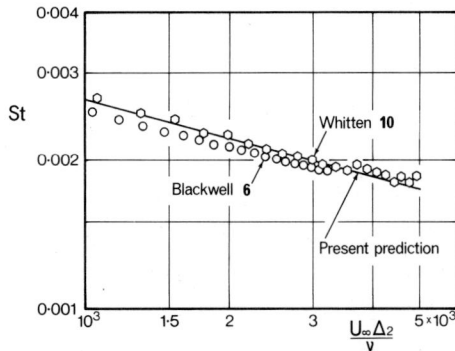

0·004

0·003

St

0·002

Whitten **10**

Blackwell **6**

Present prediction

0·001

10^3 1·5   2   3 5×10^3

$\dfrac{U_\infty \Delta_2}{\nu}$

Fig. 44   A comparison of measured and predicted values of Stanton number for a flat plate flow without transpiration.

The corresponding heat-transfer results are shown on Fig. 44 where comparison is made with two sets of data from the Stanford project, the results of Whitten [10] and of Blackwell [6]. The Blackwell data at Re_h below 2000 are a little lower than would be expected for a corresponding equilibrium thermal boundary layer, because the thermal boundary layer started out at the beginning of the test section much thinner than the momentum boundary layer.

It should also be added that all of these results were obtained using low velocity air with temperature differences from 25 to 35 degrees F. Although the influence of the temperature dependent fluid properties has not really been systematically investigated, and indeed small temperature differences were deliberately used to avoid this problem, calculations with the computer program using real properties suggest that the temperature-difference effect is to reduce the Stanton number by about 1 or 2 percent. This effect has not been considered in any of these results; the predictions have been made using constant properties, and the experimental data have not been corrected for any variable properties effect.

Note that if the friction-coefficient prediction on Fig. 43 is acceptable, the heat-transfer prediction on Fig. 44 is entirely dependent upon the distribution of turbulent Prandtl number employed, because everything else in the model is identical, and in fact both predictions were made simultaneously.

In the upper part of Fig. 45 both friction and heat transfer results are shown for a test run in an adverse pressure gradient without transpiration; the exponent m in the formula for free-stream-velocity variation being -0.15. Both the Clauser shape factor, G, and β were found to be essentially constant for the experimental data over most of the test section, so this is believed to be an equilibrium momentum boundary layer.

The prediction program also produced essentially constant values of G and β. The friction prediction

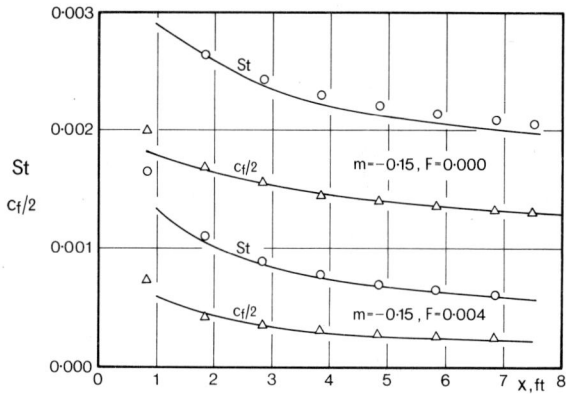

Fig. 45   A comparison of measured and predicted
            values of Stanton number and friction fac-
            tor in a mild deceleration with transpi-
            ration.

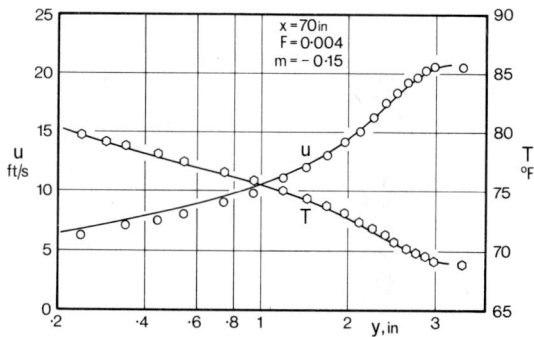

Fig. 46   A comparison of measured and predicted pro-
            files of velocity and temperature for mild
            deceleration with transpiration.

is excellent, but the heat-transfer prediction is about
5% low.   Experimental uncertainty may account for this
difference, but it is also quite possible that we see
here evidence of a pressure gradient influence on turbulent Prandtl number.   A 10% decrease of Pr_t through-

out the boundary layer would make the difference.

In the lower part of Fig. 45 are the results for an adverse pressure-gradient test with rather strong blowing. Because F was held constant (.004) this is not an equilibrium boundary layer, either momentum or thermal. For an equilibrium boundary layer it would be necessary for v_o, and thus F, to decrease with x. However, the departure from equilibrium is not great and the predictions for both $c_f/2$ and St are seen to be quite good.

The scheme described not only predicts $c_f/2$ and St quite adequately, but does equally well for velocity and temperature profiles. Figure 46 shows a pair of profiles for the adverse pressure gradient, strong blowing case discussed above. These are presented in dimensional coordinates so that normalization will not tend to mask anything, and are presented for a point 70 inches downstream so that a small percentage drift of the predictions would show as a large effect. The results shown on this figure would be hard to improve upon.

The final illustration, Fig. 47, shows an example of prediction of a very difficult case. In this run the flow starts at constant free-stream velocity but with relatively strong blowing, F = 0.004. This flow is then subjected to a very strong acceleration starting at x = 2 ft. In approximately the middle of the accelerated region the blowing is removed entirely. Then at about x = 3.4 ft, the acceleration is removed, and for the remainder of the test section there is no blowing and no change in free-stream velocity.

An important thing to note here is that the model

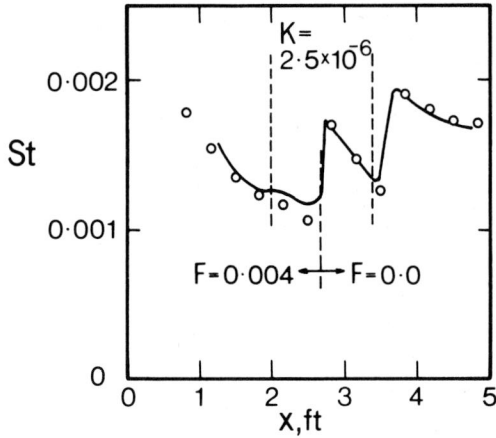

Fig. 47   A comparison of measured and predicted
             Stanton numbers for a step decrease in blow-
             ing within a strong acceleration.

responds remarkably to the abrupt changes in boundary
conditions, and predicts the resulting non-equilibrium
boundary layer very well indeed.   Of particular significance is the abrupt rise in Stanton number following the removal of blowing.   The ability of the prediction to follow the data at this point is heavily
dependent upon the use of the rate equation and lag
constant, Eq. (53).   This shows very graphically the
importance of the sublayer and the fact that the sublayer does not instantaneously assume its new equilibrium thickness after an abrupt change of boundary
conditions.   There may well be significant non-equilibrium effects in the outer part of the boundary
layer, but these have a relatively minor influence on
overall heat transfer rate.

5.   REFERENCES

1.   Mickley, H.S., Ross, R.C., Squyers, A.L. and
      Stewart, W.E.   "Heat, mass and momentum trans-
      fer for flow over a flat plate with blowing or
      suction". NACA TN 3208, 1954.

2.   Mickley, H.S. and Davis, R.S.   "Momentum transfer
      for flow over a flat plate with blowing".
      NACA TN 4017, 1957.

3.   Clauser, F.H.   "Turbulent boundary layers in ad-
      verse pressure gradients". *Jn. of Aero. Sci.*
      <u>21</u>, 91, 1954.

4.   Bradshaw, P.   "The turbulence structure of equi-
      librium boundary layers".   *J. Fl. Mech.* <u>29</u>,
      625, 1967.

5.   Andersen, P.S.   "The turbulent boundary layer on
      a porous plate: an experimental study of the
      fluid mechanics for adverse free stream press-
      ure gradients".   Ph.D. Thesis, Stanford Univer-
      sity, 1972.

6.   Blackwell, B.F.   "The turbulent boundary layer
      on a porous plate: an experimental study of
      the heat transfer behaviour with adverse press-
      ure gradients".   Ph.D. Thesis, Stanford Univer-
      sity, 1972.

7.   Moffat, R.J.   "The turbulent boundary layer on a
      porous plate: experimental heat transfer with
      uniform blowing and suction".   Ph.D. Thesis,
      Stanford University, 1967.

8.   Simpson, R.L.   "The turbulent boundary layer on
      a porous plate: an experimental study of the
      fluid dynamics with injection and suction".
      Ph.D. Thesis, Stanford University, 1967.

9.   Squire, L.C.   "The constant property turbulent
      boundary layer with injection: a reanalysis
      of some experimental results". *Int. J. Heat
      and Mass Transfer* <u>13</u>, 939, 1970.

10.   Whitten, D.G.   "The turbulent boundary layer on
      a porous plate: experimental heat transfer
      with variable suction, blowing, and surface
      temperature".   Ph.D. Thesis, Stanford Univer-
      sity, 1967.

11.   Black, T.J. and Sarnecki, A.J.   "The turbulent

boundary layer with suction or injection".
ARC R & M 3387, 1965.

12.   Stevenson, T.N.   "A law of the wall for turbu-
        lent boundary layers with suction or injec-
        tion".   Cranfield College of Aero. Report 166,
        1963.

13.   Thielbahr, W.H.   "The turbulent boundary layer:
        experimental heat transfer with blowing, suc-
        tion, and favourable pressure gradient".
        Ph.D. Thesis, Stanford University, 1969.

14.   Kearney, D.W.   "The turbulent boundary layer:
        experimental heat transfer with strong favour-
        able pressure gradients and blowing".   Ph.D.
        Thesis, Stanford University, 1970.

15.   Van Driest, E.R.   "On turbulent flow near a
        wall".   Heat Transfer and Fluid Mechanics
        Institute, 1956.

16.   Bradshaw, P.   "A note on reverse transition".
        J. Fl. Mech. $\underline{35}$, 387, 1969.

17.   Rubesin, M.W.   "An analytical estimation of the
        effect of transpiration cooling on the heat-
        transfer and skin friction characteristics of
        a compressible, turbulent boundary layer".
        NACA TN 3341, 1954.

18.   Launder, B.E. and Jones, W.P.   "On the predic-
        tion of laminarization".   Presented at the
        ARC Heat and Mass Transfer Subcommittee Meet-
        ing of April 5, 1968.

19.   Simpson, R.L., Whitten, D.G. and Moffat, R.J.
        "An experimental study of the turbulent Prandtl
        number of air with injection and suction".
        Int. J. of Heat and Mass Transfer $\underline{13}$, 125,
        1970.

20.   Spalding, D.B. and Patankar, S.V.   "Heat and
        mass transfer in boundary layers".   Morgan-
        Grampian, London, 1967.

21.   Coles, D.E.   "The turbulent boundary layer in a
        compressible fluid".   Rand Corporation Report
        R-403-PR, 1962.

22.   Orlando, A.F., Moffat, R.J. and Kays, W.M.
        "Heat transfer in turbulent flows under mild
        and strong adverse pressure gradient conditions
        for an arbitrary variation of the wall tempera-

ture".   Heat Transfer and Fluid Mechanics Institute, 1974.

6.   NOMENCLATURE

Symbol	*Meaning*
A^+	dimensionless length scale for the damping function
B_h	blowing parameter of the heat transfer problem
B_m	blowing parameter of the momentum problem
c	specific heat at constant pressure
c_f	coefficient of skin friction
c_{f_o}	coefficient of skin friction with no transpiration, other factors remaining constant
D	damping function for mixing length
F	blowing fraction, $\rho_o v_o / \rho_\infty U_\infty$
G	Clauser shape factor
G_∞	mass velocity of the free stream
H	shape factor, δ_1 / δ_2
k	mixing length constant
K	acceleration parameter
ℓ	mixing length
ℓ_T	length scale of turbulence
m	exponent describing free stream velocity variation in decelerating flows
$\dot{m}''$	transpiration rate
p	pressure
p^+	dimensionless pressure gradient, $\dfrac{\nu}{\rho U_\tau^3}\dfrac{dp}{dx}$
Pr	Prandtl number
Pr_t	turbulent Prandtl number
$\dot{q}''$	heat transfer rate

Symbol	*Meaning*
Re_h	enthalpy thickness Reynolds number, $G_\infty \Delta_2 / \nu$
Re_m	momentum thickness Reynolds number, $G_\infty \delta_2 / \nu$
Re_t	turbulence Reynolds number
St	Stanton number
St_o	Stanton number with no transpiration, other factors remaining constant
T	temperature
T_o	surface temperature
$\overline{T}$	dimensionless temperature
t^+	dimensionless temperature
t_d^+	dimensionless temperature defect
T_∞	temperature of free stream
u	time-averaged, x-direction velocity
u'	fluctuating component of x-direction velocity
$\overline{u}$	dimensionless velocity, u/U_∞
u^+	dimensionless velocity, u/U_τ
U_∞	velocity of the free stream
U_τ	shear velocity, $\sqrt{\tau_o/\rho}$
v	time-averaged y-direction velocity
v_o	velocity of the transpired fluid, at the wall
v_o^+	velocity of the transpired fluid, at the wall, dimensionless v_o/U_τ
v'	fluctuating component of y-direction velocity
x	distance in the stream-wise direction
y	distance normal to the wall
y^+	dimensionless distance from the wall, yU_τ/ν
z	distance normal to the flow, parallel to the surface

Symbol	Meaning
α	thermal diffusivity
β	pressure gradient parameter
Δ_2	enthalpy thickness
Δ_3	defect enthalpy thickness
$\delta(\ )$	uncertainty in ( )
δ_{99}	boundary layer thickness at the location where $u/U_\infty = 0.99$
δ_1	displacement thickness
δ_2	momentum thickness
δ_3	Clauser thickness
ε_h	turbulent diffusivity for heat
ε_m	turbulent diffusivity for momentum
λ	mixing-length proportionality factor
ν	kinematic viscosity
ρ_o	density evaluated at the surface
ρ_∞	density evaluated in the free stream
τ	shear stress
τ_o	shear stress at the surface
τ_t	turbulent shear stress

Subject Index